Social Media and Society

This book offers a detailed exploration of the role social media plays in our daily lives and across a variety of contexts, from social networking sites, messaging applications, and enterprise communication platforms to virtual reality.

Offering readers an introduction to the uses, effects, and central debates surrounding the subject of social media, this text is organized into three sections, each with a distinct focus. Part I provides an overview of social media, defining it through communication characteristics and exploring both theoretical and practical approaches to understanding it; Part II examines the impact of social media on individual users, including its effects on expression, health, and relationship management; and Part III investigates the wider social implications of social media, including its impact on politics, entertainment, marketing, and information consumption. Featuring key contemporary case studies and learner-centered exercises throughout, this book offers a rich and engaging look at the most pertinent issues of the social media era on both an individual and societal level.

This is an essential text for students of digital media, communication, journalism, and beyond, as well as a useful resource for researchers and industry professionals interested in exploring the social and psychological effects of social media.

Qihao Ji is Associate Professor of Communication at Marist College, USA. He teaches and conducts research on topics related to social media, media psychology, and entertainment media. His work on these subjects has appeared in various journals, including *New Media & Society*, *International Journal of Communication*, *Human Communication Research*, *Journalism and Mass Communication Quarterly*, *International Journal of Press/Politics*, and *Media Psychology*, among others.

Social Media and Society

Qihao Ji

NEW YORK AND LONDON

Designed cover image: rodwey2004 / Shutterstock

First published 2024
by Routledge
605 Third Avenue, New York, NY 10158

and by Routledge
4 Park Square, Milton Park, Abingdon, Oxon, OX14 4RN

Routledge is an imprint of the Taylor & Francis Group, an informa business

© 2024 Qihao Ji

The right of Qihao Ji to be identified as author of this work has been asserted in accordance with sections 77 and 78 of the Copyright, Designs and Patents Act 1988.

All rights reserved. No part of this book may be reprinted or reproduced or utilised in any form or by any electronic, mechanical, or other means, now known or hereafter invented, including photocopying and recording, or in any information storage or retrieval system, without permission in writing from the publishers.

Trademark notice: Product or corporate names may be trademarks or registered trademarks, and are used only for identification and explanation without intent to infringe.

Library of Congress Cataloging-in-Publication Data
Names: Ji, Qihao, 1989– author.
Title: Social media and society / Qihao Ji.
Description: New York, NY : Routledge, 2024. | Includes
 bibliographical references and index.
Identifiers: LCCN 2023018843 (print) | LCCN 2023018844 (ebook) |
 ISBN 9781032392455 (paperback) | ISBN 9781032399164
 (hardcover) | ISBN 9781003351962 (ebook)
Subjects: LCSH: Online social networks. | Social media. |
 Internet—Social aspects.
Classification: LCC HM742 .J52 2024 (print) | LCC HM742
 (ebook) | DDC 302.23/1—dc23/eng/20230421
LC record available at https://lccn.loc.gov/2023018843
LC ebook record available at https://lccn.loc.gov/2023018844

ISBN: 978-1-032-39916-4 (hbk)
ISBN: 978-1-032-39245-5 (pbk)
ISBN: 978-1-003-35196-2 (ebk)

DOI: 10.4324/9781003351962

Typeset in Sabon
by Apex CoVantage, LLC

For Wen.

Contents

Preface   *x*

PART I

1   What is social media?   3
Defining social media 5
The scope of social media 9
Changing media, changing culture 16

2   Social media users   19
Digital divide on social media 20
Psychological traits of social media users 22
Lurkers versus posters 24
Trolling and cyberbullying 26
The iGen 27
The iPhone effect 29
Motivations of social media use 30

3   Social media design and affordance   34
Affordance 37
Affordances of social media: the user's perspective 38
*The influence of platform affordances on social
   interactions 40*
*Affordances of social media: the organizational
   perspective 42*

4 Social media economy   50
Algorithm 50
Platform economy 57
The attention crisis 62

PART II

5 Social media and the self   71
Self-disclosure and self-presentation 72
Social media and identity development 77
Social media and memory 78
Selfie 79
Mourning and grieving on social media 82

6 Social media for relationship management   85
The nature of relationship 85
Stages of online relationship 86
Problematic use of social media in relationships 93
Social media in professional relationships 94

7 The use of social media among children and older adults   99
Developmental stages of children 99
YouTube for children 100
Issues about child-oriented video content 101
Child influencers 102
Virtual reality for children 106
Older adults' use of social media 108
Key Issues in older adults' use of social media 110

8 Social media for health and fitness   115
Appeals of using social media for health information 116
Issues about health-information seeking via social media 117
Trends in social media healthcare 119
Social media for fitness and exercise 122
Self-tracking 123

PART III

9 Social media for news and information sharing 133
 Social media for news use 134
 Echo chambers and filter bubbles 137
 Fake news, misinformation, and disinformation 138
 Policy and legal challenges regarding social media news consumption 143

10 Social media for social movement and political campaign 149
 Defining social movement 149
 How do social media support protests and movements 151
 Issues regarding social-media-empowered movements 155
 Social media for political campaign and election 157
 Key issues in social-media-driven campaigns and elections 162

11 Social media marketing 169
 Social-media-marketing triad 170
 Social media influencers 175
 Key issues in social-media-marketing industry 179

12 Social media entertainment and well-being 184
 Types of social media entertainment 184
 Defining entertainment 188
 How users enjoy social media entertainment 190
 From entertainment to well-being 191
 The impact of social media on well-being 193

Index 200

Preface

As a casual theatre-goer, the Antoinette Perry Award for Excellence in Broadway Theatre, commonly known as the Tony Award, is my go-to place for what to watch. In 2017, *Dear Evan Hansen* won the most awards of the season, including Best Musical. If you haven't seen it, the British newspaper *The Guardian* describes it as "A story of isolation in a hyper-linked world, it is written and performed with an emotional acuity that defies hashtag simplifications and 140-character limits."

When the classy Broadway starts to write stories about social media, one knows that the subject has entered the mainstream.

I began teaching courses about social media in 2013, a period that many industry observers consider a turning point for social network sites (due to, for instance, the wide adoption of "like" and "share" buttons). Despite being a self-proclaimed tech enthusiast back then, preparing to teach such a trendy subject turned out to be far from effortless. At the time, there was no well-established introductory book in the market, and most academic research about social media was either work-in-progress or not comprehensive enough. Fortunately, my students were unflinching, and as the class progressed, I quickly learned that they had ample stories they wanted to share and discuss. Their enthusiasm carried me through my first course on the subject.

In the subsequent decade, as I continued to teach and study social media, I became increasingly perplexed by the fact that academia as a whole has accumulated a significant amount of research on the uses and influences of social media in virtually all aspects of society. Yet, the knowledge seems to have suffered from a lack of integration, with a large number of important works being circulated within their own disciplines. As a result, we are left with a fragmented understanding of the ways in which social media shapes our lives and communities. In some ways, this is understandable because integrating such a diverse range of perspectives into an introductory book poses at least three major challenges.

First, social media are intrinsically tied to technological innovations, from basic internet infrastructures to machine learning algorithms. The volume of knowledge involved in creating functional social media can be difficult to catch up on, let alone relay to others. Moreover, new concepts and ideas are emerging rapidly. As I write this book, the notion of Web 3 is trending among the tech community, promising a more decentralized cybersphere and more freedom of information exchange. But attempts to find a proper definition for Web 3 lead to more questions than answers.

The second challenge I see in writing a book like this is that social media, as a term, is almost tantamount to the sea of information available on it. Events such as an unexpected punch on the stage of the Academy Awards ceremony, a gigantic balloon floating over the continental US, or the death of a royal family member create headlines, generate likes, shares, comments, and memes. In some sense, they are what makes social media grappling and valuable. But from a different perspective, it also means that the stories that I find heartwarming or gut-wrenching may well be drastically different from yours. There seems to be no common experience on social media.

Lastly, as academics, part of our routine business entails conjuring concepts and testing theories. However, the sheer variety of social media platforms, features, and content available out there means that no one theory or discipline will have the final say or sufficient explanatory power on everything related to social media. To scratch the surface of social media demands not just a good amount of exposure to the state-of-the-art research across multiple fields but also deep engagement with current industry practice and development.

For these and other reasons, a book about the intersections between social media and society is ambitious in its aims. Fortunately, with the burgeoning scholarship on social media in the past decade, the landscape is much clearer than ever. This book covers many topics, from industry-focused questions such as why modern social network sites look so similar, to more significant societal impact questions like whether smartphones have destroyed a generation. It also delves into topics such as how influencer marketing works and whether social bots can alter public opinion and affect election outcomes.

This book draws on the latest academic literature from communication studies, journalism, psychology, political science, business, and information/computer science. I am agnostic to the methodologies or academic paradigms and consider empirical studies based on surveys, experiments, in-depth interviews, content analyses, and computational techniques, as well as studies focusing on the political economy of social media and those taking a more critical cultural approach. The main criterion for inclusion in this book is the extent to which the writing informs readers about a given topic.

However, as a researcher trained in quantitative social science, I must acknowledge that I tend to be drawn toward studies that involve research design and causal inferences. To avoid assuming a similar background from my readers, I have attempted to explicate how these studies are being conducted in plain terms. Occasionally, I have also taken a few side trips to discuss the methodological issues observed in the literature. The aim is to show how the decisions researchers make while conducting original studies can affect the conclusions they draw.

A book on this subject will also be written on phenomenological grounds, with many of the issues or problems covered embedded in recent news events, cultural movements, and ongoing social changes. For example, despite the world seemingly having moved beyond the plague, the COVID-19 pandemic's scars are still fresh and palpable, with long-lasting consequences such as the rise of online misinformation, the work-from-home movement and workplace surveillance, and the mental health crisis among adolescents, to name a few. Much of the pandemic's impacts are manifested in how people use and approach social media. For this reason, I have decided to engage in conversations about the pandemic in many parts of the book rather than frowning upon them. My hope is that these discussions will add to society's collective effort in drawing appropriate lessons from this period in our history.

This book is not one that will solve your thorny issues while using social media; it does not purport to dispense tips on how to accumulate more followers, create viral campaigns, become successful influencers, or stir public opinions. Instead, it seeks to ask nuanced questions and explore the less explored intersections. In that sense, the book is diagnostic in its nature. It aims to intrigue rather than prescribe definitive answers.

Structurally, the book it contains 12 chapters that are divided into three parts. Part I, consisting of Chapters 1 through 4, starts with a holistic view of social media platforms. We will discuss the pluralistic meaning of "social media," what constitutes the sociality of social media, the design and affordances of various social network sites, and how the economy of social media operates. Then, in Part II, composed of Chapters 5 to 8, we will delve into the uses and effects of social media use from the users' perspectives, tapping into the profiles of average social media users, the psychological mechanisms that are oft-invoked in explaining social media use and effects, as well as how we use social media for relationship development and maintenance, and health and exercise. The scope then widens in Part III—Chapters 9 to 12—looking into the wide variety of contexts in which social media played their roles, be it about news, politics, marketing, entertainment, or well-being. We will survey some of the prominent issues pertaining to the content aspect of social media, and in doing so, channel concepts and ideas presented in the previous two parts. Needless to say, the issues that I chose to cover in these topic-specific areas are bound by the *status quo* of technological development, evolving practices of the industry, what's available in the scholarship, and my own knowledge repertoire; therefore, I will be grateful for suggestions that can make the scope of the book more comprehensive in the coming years.

In a way, researching and writing this book has been a personal journey filled with intellectual twists and turns. Now, as you travel down this path, I hope this book can serve as a guide and companion for some of the "must-sees." But most importantly, don't take my words for it; keep an open mind for new ideas and perspectives; explore every cranny and nook that excites you; and do your own research, if necessary. I am sure you will find the process worth your while.

Let me close by expressing my deep gratitude to those who have supported me in the completion of this book. My first thanks go out to my friends Jennifer Ciotta, Ling Lu, Daniel Wagner, and Simon Zhang, whose support and companionship fueled my writing. My appreciation goes as well to Arthur Raney, my academic mentor, who has been a guiding light in my career, and I am thankful for his wisdom and mentorship. An entire generation of students at Marist College and Florida State University who took my courses have been a constant source of inspiration, and their enthusiasm has taught me much. I profited from my colleagues at Marist College, both past and present, including Kevin Lerner, Jen Eden, Wenjing Xie, Lyn Lepre, and Jacqueline Reich, who provided me sufficient mental space to start and finish this work. In the final phase of the project, I also benefited from Troy Phay's copyediting and assistance in organizing the manuscript. I am thankful to my Routledge editors, Sheni Kruger, Emma Sherriff, and Grace Kennedy, for their help throughout the publication process. Finally, I want to express my deepest love and gratitude to my family, especially my talented and vivacious wife Wen, to whom this book is dedicated.

Part I

1 What is social media?

Here's a thought experiment: If an omnipotent figure were to bestow upon you the opportunity to choose any time period in human history to spend your life in, which period would you choose? Whatever answer you may give, there is only one rational response: you should want to forgo such an opportunity and stay in the present. The reason is that, as Harvard cognitive psychologist Steven Pinker vividly presents in his book *Enlightenment Now*, the world has been improving (on average) in almost every crucial aspect that a society should care about. There have been increases in food availability, literacy, education, leisure time, and life expectancy, and decreases in war, poverty, child mortality, sexism, and homophobia. (Let's all agree to put the years of COVID-19 aside for a while.)

But let's bring it closer to home. As an average citizen living in the second decade of the 21st century, think about the convenience and choices we have at our fingertips. With just a few taps on our smartphones, we can communicate with loved ones, share travel photos, stay informed on current events, collaborate with colleagues, find a good laugh during a break, and even pursue our hobbies. There has never been a time in human history where the average person's life was this full of possibilities and opportunities. And that's not even considering the impact of social media on our daily lives. Truly, we are living in an exciting and productive era.

Although the idea of social media is deeply ingrained in the creation of the first prototype of the Internet, namely the ARPANET (Advanced Research Projects Agency Network), the term "social media" was not officially used in print until 1997. Writing somewhat prophetically in the business context, the then AOL executive Ted Leonsis noted that corporations need to provide consumers with "social media, places where they can be entertained, communicate, and participate in a social environment" (as cited in Bercovici, 2010). Coincidently, that same year, SixDegrees.com—the first popular social network site which allows people to create online profiles and friend lists—was launched. The following decades saw blogs, Wikipedia, Myspace, Friendster, Facebook, and Twitter follow suit,

4 Part I

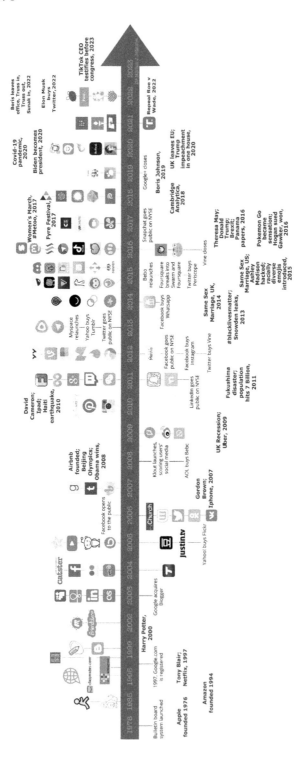

Figure 1.1 Social Media Timeline and major world events

(*Source*: Illustration by Dr. Miriam J. Johnson, reproduced with permission)

generating waves of trends in the cybersphere. And most recently, the wide availability of smartphones and high-speed Internet made what we now understand as social media part of a daily necessity.

Notwithstanding the abundance of social media, the implications of these emerging communication technologies for individuals and society are heatedly contested both within academia and among the general public. One needs not to venture too far from the popular press to find euphoric proclamations such as "How social media can revolutionize your business" to the alarm-sounding assertions as in "Yes, social media really is undermining democracy." But before we can assess these statements in any serious measure, it is helpful to discuss a little bit about what we mean by the term "social media."

Defining social media

Typology of social media

One approach to understanding the definition of social media is to typologize a list of commonly used sites and services. In that view, social media must include social network sites: platforms and services designed to facilitate social interactions among people in various settings. Hence, sites such as Facebook, LinkedIn, Nextdoor, and Tinder appear to be intuitive examples. However, depending on the degree to which "sociality" or "socialness" is manifested within a given service, many users also consider social media to entail video sharing sites (TikTok), game streaming sites (Twitch), messaging services (Discord), microblogs (Twitter/X), Internet forums/discussion sites (Reddit), gig economy apps (Airbnb), review sites (TripAdvisor), collaboration tools (Slack), and bookmarking/hobby gathering sites (Pinterest).

Exercise

Given your extensive experience as a social media user, can you come up with a few more examples under each type of social media listed above? Do you think that this typology of social media is complete? If not, what other types of social media may you add? Now, consider this: with so many types of social media available in the market nowadays, what makes social media social? In other words, where is that "sociality" coming from within specific social media platforms and across the broad?

Social media scholar Christian Fuchs (2017) listed four possible dimensions of sociality in social media: information, communication, communities, and collaboration. How do you interpret these four dimensions of sociality? Do you think that this framework sufficiently captures the socialness of social media?

Such a typology of social media is undeniably useful in that it offers a quick and easy way to understand various platforms or services in terms of their *defining* functions. But there are drawbacks: for one, it is sometimes difficult to clearly demarcate one type of platform from another because users may not have agreed views as to the main utile of a site or an app. For example, Snapchat is seen by many as an app for "photo/video" sharing (presumably because of its visual messaging features), yet a significant number of users also see it as a "social" site due to its functionalities for spontaneous social interaction (Rhee et al., 2021).

And then there is the issue of technological improvement. As nearly all self-identified social media users can attest, no platform or service will stay static during its lifetime; rather, new features are constantly being tested and added on to the platform, while the old ones are being modified or eliminated. In this process, the category under which a platform was initially conceived may no longer capture what it now resembles. Take YouTube for example: it was a platform initially designed for online video sharing, but with the inclusion of features such as YouTube community, YouTube Stories, and YouTube Shorts over the years, the site has since become something a lot more like a mainstream social network site. And continuously referring to YouTube as a video-sharing platform misses its true value among its users. For these reasons, conceptualizing social media through a list of typologies can easily lead to confusion and miscommunication.

The communication characteristics of social media

If the folk definition of social media in terms of typology is not the best route, are there other approaches for defining social media? Enter the scientific approach.

Historically, the communication field has distinguished between mass communication and interpersonal communication, with the prior defined as (a) one-way, (b) technologically mediated messages, (c) delivered to large audiences (d) of individuals not known personally by the sender (think, televisions). And the latter defined as (a) two-way, (b) non-mediated message exchange between (c) a very small number of (usually two) participants (d) who have personal knowledge of each other (e.g., face-to-face chat). Obviously, this division becomes less instructive as social media join the scene. To illustrate, a seemingly mass-communication-oriented site such as the game streaming platform Twitch can be used for semi-interpersonal conversation, as shown in its popular content category Just Chatting (i.e., streamers broadcast themselves engaging in non-gaming-related activities, such as cooking, walking around town, or just sitting in front of their camera talking to viewers). Conversely, a somewhat interpersonal app such as the instant messaging service Telegram are simultaneously used by many to broadcast news, propaganda, and entertainment.

To properly define social media and link the increasingly outdated dichotomy between interpersonal and mass communication, some scholars (O'Sullivan & Carr, 2018) proposed to conceive various forms of communication technologies and channels by two dimensions: (a) perceived message accessibility, and (b) message personalization. Perceived accessibility denotes the exclusivity of message access as perceived by the message creator. In other words, communication channels may vary in terms of the number of intended audiences, ranging from a single person to any maximum number of audiences a technology may afford to reach. On the other hand, communication channels vary by message personalization, which can be understood as the degree to which receivers perceive a message reflects their distinctiveness as individuals differentiated by their interests, history, relationship network, and so on. As such, communication channels may be evaluated on a continuum anchored by highly impersonal and highly personal.

Together with these two dimensions, one can compartmentalize many of the familiar social media platforms (and specific features) under four distinctive quadrants (see Figure 1.2): quadrant I entails the highly tailored but rather public communication channels such as Twitter mentions, Facebook comments and likes, and personalized celebrity videos from Cameo. Quadrant II comprises the highly tailored social media channels which

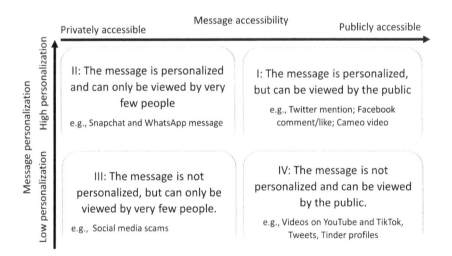

Figure 1.2 Social media quadrant

(*Source*: Figure redrawn based on O'Sullivan & Carr, 2018)

are typically accessed by very few people. Familiar examples include, for instance, various instant messaging features of mainstream social media platforms. Quadrant III consists of messages that are not personalized, yet they can only be accessed by very few targeted recipients. Most social media scams dovetail neatly with the definition of the quadrant. Finally, quadrant IV encompasses social media channels which generate non-personalized messages that can be viewed by the general public. Examples include videos on YouTube and TikTok, Tweets, and personal profiles on dating apps.

Two more communication characteristics of social media are worth noting; namely, *media modality* and *media synchronicity*. Media content is created in specific modals or forms, be it sounds, images, texts, or videos. But unlike content generated in traditional media, such as in books, TVs, and radios, social media content is by nature multi-modal, containing a mixture of all-existing forms we already know as well as ones that may yet to fully realize (e.g., virtual reality and augmented reality). That richness of the information on social media, in terms of content modality, makes social media an apt channel for communication in ever-changing circumstances.

Lastly, we should not forget the (a)synchronicity of social media—the ability to react to communicative requests in an immediate or time-delayed manner. Understandably, synchronous communication necessitates that at least people involved in communicative settings are simultaneously online and inputting. By contrast, asynchronous communication allows temporal asymmetries of input from the involved parties, making room for more sophisticated encoding and decoding of messages. What this means, then, is that social media users can negotiate their terms in communication based on their privacy and communicative efficiency needs (Taipale, 2016). In the following chapters, we will explore these ideas in greater detail.

One particular advantage of seeing social media through the colored glasses of these communication characteristics and features is that it fosters a sophisticated understanding that social media is not just one platform, one type of service, or one specific form of technology but an evolving way of communication that varies on continua. After all, every decade has its own Myspace, Facebook, and TikTok, and as users, we ought to hold a more dynamic and nuanced view of what we see as social media. In that light, this book proposes a working definition on social media as follows:

> Social media are digital media services that enable users to consume, produce, and/or interact with user-generated content of various modalities so as to articulate meaningful connections.

> **Exercise**
>
> Some scholars argue that all media are social, unless people use them in nonsocial ways. By this train of thought, any platform or service can be considered social media. The definition of social media offered in this book seeks to provide a lens to conceive social media both from a theoretical and a functional perspective. Under this definition, can you see what is *not* social media? Provide a few examples to substantiate your argument. Finally, do you have an alternative way of defining social media?

The scope of social media

There are a multitude of reasons why we ought to care about social media. (And this book is devoted to answering this very question.) But fundamentally, all of our interests in social media stem from its materialistic nature; namely, the technology, the people, and the content. The following section will briefly introduce the critical constituents in each of these modules, as well as highlighting some of the ways in which these components become entangled.

Technology

Let's first start with the technology of social media. It is useful to broadly distinguish two types of technological foundations of social media: the ones users can see and access directly and the ones that are exclusively accessible to platform developers or researchers. For average social media users, the interface of a given platform is the technology with which we interact directly to forge interactions and create content. Here, user interface (UI) or user experience (UX) includes everything from the aesthetics such as the color of an icon, the frame of text, or the layout of content, to major platform features such as sharing photos, editing videos, and adding filters. Increasingly, though, social media UI/UX designs are mixed with both aesthetic values and functional values. Take the now-ubiquitous "like button," introduced by Facebook in 2009, for instance: it was designed literally as a mechanical button that users have to press on to activate its symbolic function—to express a favorable attitude. By transforming the amorphous digital code into something seemingly solid and familiar, we as users are presented the potential to interact with the platform and others (Pold, 2008).

But how does a social media company decide which part of their UI/UX design gets tweaked or updated? The answer is the *A/B testing*. Also known as bucket testing or split testing in the tech sector, A/B testing is an extremely common way for software developers to assess users' responses to a new feature, service, or design (Siroker & Koomen, 2013). In practice, the method is akin to scientific controlled experiments, where developers can show different design variations to large groups of users and gauge their reactions. Today, because of the digital nature of social media, which permits non-intrusive data collection, A/B testing is constantly being performed, and all the while, unbeknownst to average platform users like you and me.

On the opposite side of the visible technologies of social media, there exists a group of invisible technologies that are mostly accessible by software developers. These include algorithms, content management system, data storage, and computing, to name just a few. Again, most of these technologies cannot be directly experienced by average users, but they are nevertheless critical to the daily operation of a platform. For instance, on October 4, 2021, Facebook and its subsidiary apps (Instagram, WhatsApp, Messenger, and Oculus/Quest) experienced a global outage that lasted for nearly 5 hours due to malfunctions in crucial Internet infrastructure that coordinates the traffic between its data centers (Janardhan, 2021). Considering Meta's over 3.5 billion user bases worldwide, the social and economic impact of such an incident is not to be underestimated.

Among the invisible technologies social media comprise, two mediatory ones are uniquely relevant to users' experiences: algorithms and application programming interfaces (APIs). Briefly, algorithms are mathematical models that are performed in a controlled fashion on data to present output in other desirable forms. In the context of social media, they contribute directly to users' experiences by serving functions such as content curation and recommendation, content moderation, and advertisement allocation. (More on that in Chapter 4.) On the other hand, APIs are "interfaces that facilitate the controlled access to the functionality and data contained by a software service or program" (Bucher, 2013, p. 3). Simply put, APIs are controlled gateways through which apps and software can communicate with each other. Today, social media companies commonly offer APIs, allowing third-party developers to scrape data, access specific functionalities, and build new applications on top of the existing platforms. However, this norm is quickly changing, with companies such as Reddit and Twitter starting to impose hefty fees for the use of their APIs.

Industry participants

Beyond the technological foundations, social media would not exist without the people involved in their creation and use. However, the term "industry participants" encompasses a variety of actors who serve vastly different roles. Generally, social media is of most concern to those with the following affiliations: government regulatory bodies, service providers, and average users.

First, people in the regulatory bodies, presumably, are the ones who can exercise their rights to implement rules and regulations over social media companies and platforms under the respective country's legal framework. In the US, The Federal Communications Commission (FCC) and Federal Trade Commission (FTC) are the two primary federal agencies charged with such duties. In the European Union region, the European Parliament and Council is responsible for the most formidable legal framework, named The General Data Protection Regulation (GDPR). Under GDPR, each member state is instructed to appoint a Data Protection Authority (DPA) that is responsible for monitoring and enforcing the law. Though, depending on each EU member's federal structures, a country may have multiple DPAs. For example, in Germany, the federal DPA (Der Bundesbeauftragte für den Datenschutz und die Informationsfreiheit, or BfDI) collaborates with its 17 state DPAs to enforce the GDPR.

> Not a media company?
>
> In the US, the Federal Communications Commission (FCC) theoretically holds regulatory authority over social media companies. But they rarely do so in reality. This is, in large part, due to the fact that many social media companies are reluctant to be called a "media company." The CEOs of large social media and tech companies regularly cite one or more of the following reasons to defend their non-media attribute (Napoli & Caplan, 2017): a) "We don't produce content"; that is, social media companies are only distributors of content created by users; b) "We're computer scientists": the founders, CEOs, and a large proportion of the workforce in these companies are trained as computer/software engineers; and c) "No human editorial intervention": content seen and consumed by users on social media is decided by the system (algorithms), not by any specific individual in the company.
>
> The real benefit for these social media companies arguing against being seen as media companies (or, in legal terms, "information content providers") is such that their platforms and services could enjoy much more leeway in their operations under Section 230 of the Telecommunications Act of 1996. Broadly speaking, section 230 shelters Internet companies (including Internet service providers and social media companies) from legal liabilities due to any "third-party"

> speech happening on their platforms. For this reason, recent years have seen more and more activists, NGOs, legal scholars, and politicians advocate for amending or updating section 230. One change did happen in 2018 with the pass of the Fight Online Sex Trafficking Act (FOSTA), which made an exception out of section 230's liability shield for Internet companies' violation of federal sex trafficking law. But the debate over section 230 is very much alive today as social media misinformation gets much more pronounced. More on the matter in Chapter 9.

The second group of industry participants that social media concern the most are the service providers. Service providers refer to the ones who design, maintain, and operate the social media platforms. From an organizational point of view, individuals involved in the service providers can be crudely categorized under two types: corporate executives and software engineers/developers (including other operational personnel). Evidently, as leaders in their organizations, business executives (including the company's founders) hold significant power in influencing an organization's long-term trajectories, relations with stakeholders, reputation, among others. This is particularly the case for social media companies, where many chief executive officers are often seen as the highest authorities of respective platforms in the eyes of the general public and news media. And for this reason, congressional hearings featuring Mark Zuckerberg, Sundar Pichai, and Shou Zi Chew tend to be closely monitored.

In contrast to the business executives, software engineers/developers appear to be much less flamboyant. Generally speaking, there are three main types of software engineers in most social media companies: back-end, front-end, and full-stack engineers. Back-end engineers are responsible for the server-side of the application, which includes designing and developing the back-end systems such as databases, servers, and APIs. On the other hand, front-end engineers focus on the user-facing part of the application, which includes the development and design of the user interface. Full-stack engineers, as the name suggests, have a combination of back-end and front-end duties and work on both the server side and client side of the application. They can understand and implement the complete web development process, from the back end to the front end.

By and large, these three types of engineers form the backbone of any social media company, but there are also other types of engineers that are task-specific, such as mobile engineers, data engineers, infrastructure engineers, security engineers, and machine-learning engineers, to mention a few.

Together, software engineers ensure the smooth functioning and performance of the product social media services, addressing issues as little as fixing a minor bug in a line of code to the ones as big as the lunch of a new feature. Arguably, software engineers are the most integral members among the service providers, yet their organizational roles as corporate employees also make them prone to fall victim to exploitation and burnout. When Twitter was sold to Elon Musk in late 2022, engineers of the company received direct communication from the business mogul to work "long hours at high intensity" (Kolodny, 2022). From this perspective, software developers are sometimes seen as "digital laborers," just as those social media users who generate values through content creations (Fuchs & Sevignani, 2013).

The third group of industry participants that social media concern the most are the users of social media. As we will soon discover in Chapter 2, sketching out a typical profile of an average social media user turns out to be a nearly impossible task; for the sheer variety of platforms available in the market nowadays cover almost the full spectrum of users both in terms of demographics and social economic status. Nevertheless, changes in the demographic compositions of the general population will have profound social and economic impacts. Notably, for instance, with a population of over 62 million, Latinx is among the fastest growing racial/ethnic groups in the United States, with a medium age of just hitting 30 (the national medium for Asian, Black, and White are 38, 35, and 44, respectively; Noe-Bustamante et al., 2020). As such, the Latinx group is seen as a fresh propulsion to the media and entertainment industries (Nilsen, 2021).

Also, in an increasingly user-centered social media economy, users are no longer just consumers but are also content creators. This juxtaposition of consumer and creator identities means that average users' needs and desires, self-expressions, and artistic creations are becoming more important than ever. In that sense, social media users, in and of themselves, are a powerful source for political and cultural change. In the meantime, however, users are also feeding "raw materials" to the commercialized social media companies in three forms of data: public data, private (or semi-public) data, and meta-data (Fisher, 2015). Public data mainly refers to user-generated content, such as videos on YouTube, photos on Instagram, and tweets on Twitter. Private data refers to the information generated through users' interaction with the platforms and other users, which consists of search queries, likes, posts, shares, and others. And meta-data entails the digital footprint generated by merely using the social media services, such as our location, the type of our device, frequency of a specific, number of friends on the network, how long we spent on each piece of content, and so on. Together, these data contributed to a highly sophisticated and technical social-media-analytics industry. To the eyes of digital analysts, user information is the input, and the predicted user behaviors are the output.

Content

Ultimately, what interests and matters the most to individuals, industries, academics, governments, and societies at large is the content generated by users of various kinds. In the vast landscape of social media, content is neither homogeneous nor monolithic. It is human motivations infused with pluralistic ideologies, which is then presented in countless modalities. That means the meaning-making of social media content, however is done, should not be devoid of specific context, be it business, health, journalism, entertainment, or politics, among others. Part III of the book will dive into these contextual settings in greater detail.

But there is a commonality in how content is produced on social media. And that has a lot to do with the popular business lingo "Web 2.0." The term was coined by Tim O'Reilly (2005) in the early 2000s to describe a new type of Web that is distinctive to the old and prior Web 1.0, which features a network of static hypertexts with limited user interactions. What the Web 2.0 era promised is a new "architecture of participation" (O'Reilly, 2005, para. 24), which facilitates the co-production of content, user interactivity, and social networking. And many of the platforms that people are now familiar with—Facebook, Instagram, Snapchat, Twitter, and TikTok—are tools that actualize Web 2.0's capacities for user-generated content. As of this writing, there is a rapidly evolving conversation (largely happening among the business sector) about the potentials of the Web 3.0, a so-called decentralized Web. A consensus about what the term Web 3.0 means and entails is far from being reached. One thing that is certain, though, is that users will always be at the front row to witness and experience the change in the way they produce and distribute content.

Web 3.0

Before the business and tech community made the term "Web 3.0" (or Web 3) an Internet buzzword, Tim Berners-Lee, the creator of the World Wide Web, first coined the term in 1999 to envision a next generation of the Internet where machines could communicate with one another and to understand and create meaning from semantic data (i.e., Semantic Web). Many scholars suggest that, in contrast to the current Web 2.0, which primarily features user participation, Web 3.0 is thought to be featured by users' cooperation and co-ownership. To illustrate the collaborative nature of Web 3.0, Barassi and Treré (2012) invoke the example of Quora, a popular social question and answering website in which users co-own and co-create questions

> and answers posted on the site. In reality, though, Web 3.0 and its applications in the social media arena is still a work in progress, and the conflation of Web 3.0 with blockchain technology, cryptocurrency, and non-fungible tokens adds even more layers of confusion (Stackpole, 2022). However, as you are reading this chapter, there might be some platforms out there that are built on the premise of Web 3.0. Try to find a couple of those platforms and play around with it yourself.

Four core issues about social media content

Regardless of the technological change and the era it defines, there will be a few common challenges that permeate the content generated online and the social reality. In that regard, danah boyd, one of the pioneers in the field of social media research, foresees four core issues (boyd, 2010) that still apply in today's context: democratization, stimulation, homophily, and power.

The first issue, democratization, refers to the problem that, as social media democratizes the access to information (and, to a large extent, the production of information), attention to information does not get divided equally. In other words, there's no guarantee that what people pay attention to will be the most urgent or even the most meaningful. Related, the second issue, stimulation, refers to the situation where only the most sensational content online gets traction. And in the long run, content that is of social importance but otherwise lacks eye-catching ingredients may get sidelined. Thanks to social media data mining and analytics, more and more evidence supporting the democratization and stimulation trends are surfacing. For example, according to Facebook's own research, nearly three quarters of the viral content on its platform is of low quality and contains spammy or oversexualized messages (Horwitz, 2022).

The third issue that looms large on social media is homophily. Homophily refers to the phenomenon where users in an open network group together into small segregated circles, resulting in another formal of information and ideological polarization that may solidify social division. Lastly, the issue of power concerns who has the influence to command attention and generate traffic through information. In the old times, the creators or gatekeepers of the content (e.g., journalists, broadcasters, film directors, etc.) were often thought to hold such power and benefit from it, but increasingly, that power has shifted to the so-called "information brokers," those individuals and originations specialize in getting others' attention by re-distributing content rather than authoring original content.

As we will soon see, these four issues are recurring themes, and we will explicate how they are manifested in specific domains and settings in the coming chapters.

Changing media, changing culture

We started this chapter with a thought experiment on a grand time-scale, so let's close the chapter on a similar note. In the long arch of our history, humans have lived under very humble conditions for the majority of the time. Our material and mental well-being, as measured by the word's gross domestic product, was almost stagnant for thousands of years (close to zero, actually; see Figure 1.3). The real change happened only roughly in the last 100 years when our species' productivity suddenly skyrocketed. What changed? After all, we still possess the same physical features (including structures of the brain) as our ancestors did back in the day. In a 2015 TED talk based on his highly regarded book *Sapiens: A Brief History of Humankind*, Israeli historian Yuval Noah Harari attempted to explain:

> The real difference between humans and all other animals is not on the individual level; it's on the collective level. Humans control the planet because they are the only animals that can cooperate both flexibly and in very large numbers.

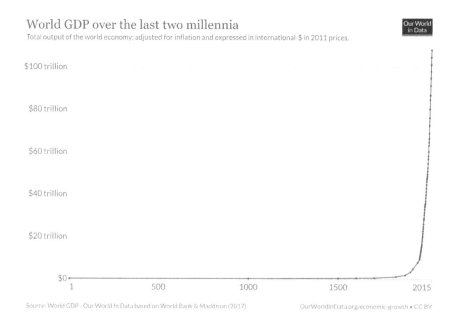

Figure 1.3 World GDP over the last two millennia

But then, what organizes people to cooperate and lead to societal change? Many would point to culture. There is a lively debate within academia on the precise definition of culture, but in simple terms, culture is ideas that make people behave alike (Deutsch, 2011). In practice, though, culture is arts, literature, music, films, food, traditions, rituals, and everything in between. Change happens when enough people engaging in these sorts of things think and do differently—we probably don't need hard evidence on this, but just for the sake of completeness: in a study published in *Science*, researchers conducted simulations using sophisticated computer models on how minority groups can initiate social change through the emergence of new social conventions. What they found, after over a thousand modeling trials, is that, when roughly 25% of the population thinks and approaches things differently, social change, such as views on gender role beliefs, racial equity, and LGBTQ+ rights, may soon follow. Needless to say, information communication technologies such as social media will be a critical imperative; as the authors (Centola et al., 2018) suggest, "we anticipate that social media spaces . . . will be an increasingly important setting for . . . changing social expectations regarding (i) the standards of civility on Facebook and other online discussion forums, (ii) the acceptability of bullying behavior in adolescent chat groups, and (iii) the appropriate kinds of content to share with strangers over social media" (p. 3).

Readers of this book (mostly Gen Zs and Gen Alphas, I assume) are certainly the core force of any changes yet to come. Through your participation and formation in various digitally mediated culture practices such as the gaming culture, mobile culture, selfie culture, influencer culture, dating culture, liking culture, meme culture, and many more, societal changes will accelerate in the coming decades.

Keywords: sociality; perceived accessibility; message personalization; synchronicity; multimodality; A/B testing; API; UI/X design; software engineers; three forms of user data; section 230; GDPR; Web 2.0; culture.

References

Barassi, V., & Treré, E. (2012). Does web 3.0 come after web 2.0? deconstructing theoretical assumptions through practice. *New Media & Society*, 14(8), 1269–1285. https://doi.org/10.1177/1461444812445878

Bercovici, J. (2010). Who coined 'social media'? Web pioneers compete for credit. *Forbes.com*. www.forbes.com/sites/jeffbercovici/2010/12/09/who-coined-social-media-web-pioneers-compete-for-credit/?sh=1d669d0f51d5

boyd, d. (2010). Streams of content, limited attention: The flow of information through social media. *Educause Review*, 45(5), 26.

Bucher, T. (2013). Objects of intense feeling: The case of the Twitter API. *Computational Culture*, (3). http://computationalculture.net/objects-of-intense-feeling-the-case-of-the-twitter-api/

Centola, D., Becker, J., Brackbill, D., & Baronchelli, A. (2018). Experimental evidence for tipping points in social convention. *Science, 360*(6393), 1116–1119. https://doi.org/10.1126/science.aas8827

Deutsch, D. (2011). *The beginning of infinity: Explanations that transform the world*. Penguin.

Fisher, E. (2015). 'You media': Audiencing as marketing in social media. *Media, Culture & Society, 37*(1), 50–67. https://doi.org/10.1177/0163443714549088

Fuchs, C. (2017). *Social media: A critical introduction* (2nd ed.). SAGE Publications. https://doi.org/10.4135/9781446270066

Fuchs, C., & Sevignani, S. (2013). What is digital labour? What is digital work? What's their difference? And why do these questions matter for understanding social media? *tripleC, 11*(2), 237–293. https://doi.org/10.31269/triplec.v11i2.461

Horwitz, J. (2022). Facebook's most popular posts were trash. Here is how it cleaned up. *WSJ.com*. www.wsj.com/articles/facebooks-most-popular-posts-were-trash-here-is-how-it-cleaned-up-11669140034

Janardhan, S. (2021). Update about the October 4th outage. https://engineering.fb.com/2021/10/04/networking-traffic/outage

Kolodny, L. (2022). Elon Musk demands Twitter staff commit to 'long hours' or leave: Read the email. *CNBC.com*. www.cnbc.com/2022/11/16/elon-musk-demands-twitter-staff-commit-to-long-hours-or-leave.html

Napoli, P., & Caplan, R. (2017). Why media companies insist they're not media companies, why they're wrong, and why it matters. *First Monday*. https://doi.org/10.5210/fm.v22i5.7051

Nilsen. (2021). Inclusion, information, and intersection the truth about connecting with U.S. Latinos. *nielsen.com*. www.nielsen.com/wp-content/uploads/sites/3/2021/09/nielsen-2021-hispanic-diverse-insights-report-210682-D9.pdf

Noe-Bustamante, L., Lopez, M. H., & Krogstad, J. M. (2020, July 7). U.S. Hispanic population surpassed 60 million in 2019, but growth has slowed. *Pew Research Center*. www.pewresearch.org/fact tank/2020/07/07/u-s-hispanic-population-surpassed-60-million-in-2019-but-growth-has-slowed/

O'Reilly, T. (2005). *What is web 2.0: Design patterns and business models for the next generation of software*. http://oreilly.com/web2/archive/what-is-web-20.html

O'Sullivan, P. B., & Carr, C. T. (2018). Masspersonal communication: A model bridging the mass-interpersonal divide. *New Media & Society, 20*(3), 1161–1180. https://doi.org/10.1177/1461444816686104

Pold, S. (2008). Button. In M. Fuller (Ed.), *Software studies. A lexicon* (pp. 31–36). MIT Press. https://doi.org/10.7551/mitpress/9780262062749.003.0004

Rhee, L., Bayer, J. B., Lee, D. S., & Kuru, O. (2021). Social by definition: How users define social platforms and why it matters. *Telematics and Informatics, 59*, 101538. https://doi.org/10.1016/j.tele.2020.101538

Siroker, D., & Koomen, P. (2013). *A/B testing: The most powerful way to turn clicks into customers*. John Wiley & Sons. https://doi.org/10.1002/9781119176459m

Stackpole, T. (2022). What's web 3? *Harvard Business Review*. https://hbr.org/2022/05/what-is-web3

Taipale, S. (2016). Synchronicity matters: Defining the characteristics of digital generations. *Information, Communication & Society, 19*(1), 80–94. https://doi.org/10.1080/1369118X.2015.1093528

2 Social media users

Among the abundance of psychology literature on human cognitive bias, there exists the notion of heuristic bias, which refers to a mental shortcut that relies on immediate examples that come to a given person's mind when evaluating a specific topic, concept, method, or decision. Active Twitter users may recall that, in early 2022, an overwhelming number of users started to tweet about a six-by-four square grid colored with yellow, green, and gray blocks. To bystanders of Internet culture like myself, I was mesmerized, only to discover that it was a crossword-like web game called Wordle, which was meant to be played once daily. To impress my tech-savvy and culturally sophisticated undergraduate students, I brought up this game in my social media class, thinking it would easily connect with the audience. To my surprise, however, students looked at me with confused eyes—none of them (except for one) had ever heard about it. Why?

Every few years, the Pew research center publishes a set of national surveys about Americans' social media use trends; here are some noteworthy facts in their 2021 report:

- 95% of Americans use YouTube, making it the most popular social media platform.
- Most Americans who are 18 to 24 years old are on Instagram (76%), Snapchat (75%), or TikTok (55%).
- About half of Hispanic (52%) and Black Americans (49%) use Instagram, compared with smaller shares of White Americans (35%) using the service.
- 33% of adults with a bachelor's or advanced degree are on Twitter, whereas 23% and 19% of them are on Snapchat and TikTok.

Examining general statistics like these can be very revealing: it turns out what we commonly refer to as social media are, in fact, completely different eco-systems with drastically different user profiles. Some platforms are particularly trendy among teens and adolescents; others are disproportionally

DOI: 10.4324/9781003351962-3

Figure 2.1 Trolling and cyberbullying on social media
(*Source*: HappyPictures/Shutterstock)

used by adults and the general population. No wonder the Wordle storm, which took over my Twitter feed, was completely unheard of among my young adult students. And that was my heuristic bias: somehow, I, too, thought that everyone was on Twitter.

Digital divide on social media

Social media users are individuals who regularly use social media platforms to connect with others, share content, and participate in online communities. These groups of people are naturally distinguished from those don't currently use any social media services at all, whether willingly or unwillingly. And such a gap between those who have access and those who do not is referred to as the digital divide (Van Deursen & Van Dijk, 2014).

Many demographic factors, such as education, income, and race/ethnicity, may account for the existence of the digital divide. But one prominent factor, on a population level, is the generation difference. By nature, young people are attuned to the rapid change of technologies and are more aware of the various new digital services that are available; hence, they are often being referred to as "digital natives," or the "digital rich." In contrast, older adults who are less sensitive to technologies and only learn to use some of the new digital services are often labeled as "digital immigrants" (Prensky, 2001); hence, digitally poor.

The generation-driven digital divide exists not only in the general population but among people of the most converted occupations. In an intriguing research, Whittaker and colleagues (2020) conducted a series of in-depth interviews and focus groups with members of street gangs in London, particularly regarding their uses of social media to sell drugs, recruit members, and control territories. Based on their analysis of interview data, the researchers identified two groups. The first, dubbed "digitalists," comprises gang members who prefer to embrace technology as a means of conducting business and developing the gang's identity and as a way to gain reputation and territorial expansion. The second group consists of those who tend to keep a distance with technology, either purposely or inadvertently; these are referred to as "traditionalists." Right there, the researchers noticed that the two groups of gang members are naturally distinguished by generational differences, with the young members gravitating more toward using social media for organized crime. Interestingly, though, old gang members' resistance to new communication technologies isn't completely tied to the lack of digital proficiency. Rather, as one became more and more criminally involved, some old gang members developed to be more vigilant and tended to avoid attracting unwanted attention.

Increasingly, however, the digital divide is less about the disparities between those who have access and those who do not but more about who has what access and of which kind. For instance, in recent years, as natural disasters became much more frequent and severe, government authorities have started to adopt social media as a key channel for disaster management, such as distributing information about weather prediction, market/food availability, road condition, and other service status. Depending on which specific platform government agencies decide to use, citizens in the impacted region may or may not have access to disaster-related information at critical moments, which are crucial to reducing economic and personal-life losses. To better understand social media use among different sub-populations and the extent to which inequalities exist for vulnerable populations in disaster situations, researchers (Dargin et al., 2021) surveyed residents in several Southern US States impacted by Hurricane Harvey, Florence, and Michael (all happened between 2017 and 2018). The study found that socioeconomic factors such as income, education, and race, along with one's geographic location (urban vs. rural) all affected people's choice of social media platforms as well as their information seeking and sharing activities during disaster periods. In particular, households in rural areas, lower income groups, and racial minorities were more likely to report greater distrust toward social media information. In other words, as access to technology and Internet improves in the society, the digital divide nowadays implies more about a divide of information access.

What the divide of information access shows, ultimately, is a form of inequality in terms of access to digital skills and usage. In that regard, individuals' education level explains quite a lot. Through a representative survey of the Dutch population, Van Deursen and Van Dijk (2014) found that "people with a low level of education use the Internet more frequently and for more hours a day than people with medium and high levels of education. Low educated people seem to engage more in social interaction and gaming, which both are very time-consuming activities" (p. 520). Essentially, this and other similar studies found that the Internet (and the digital services it supports) alone cannot remove the existing social, economic, and cultural disparities deeply ingrained in our society. In fact, as the entire population moves to the digital sphere, those who are less educated and less well-off are more likely to fall into the trap of using them primarily for fun. In contrast, the well-educated and well-off ones are more likely to use the Internet and social media for information and personal improvement. This point is very well illustrated in a conversation between the comedian and podcaster Joe Rogan and Mark Zuckerberg (Rogan, 2022, 55:00):

Rogan: Do you limit your social media use? How do you do it?
Zuckerberg: Me personally, I am just doing so many things, so I am not [using social media]. . . . Everything you are doing on a computer or a screen isn't the same. . . . If you are just sitting there and consuming stuff, I mean, it's not necessarily bad, but it's generally is not associated with all the positive benefits from being actively engaged and building relationships.

Psychological traits of social media users

Our discussion about the profiles of average social media users thus far has been centered around demographic factors. However, as instructive as they could be, some may find these demographic factors too broad to capture the subtle differences across individual users. As a result, researchers and practitioners have turned to the vast collection of psychology literature for potential answers. In particular, many have looked at the connections between individuals' personality traits and social media use under the well-established Big Five framework. Generally speaking, the Big Five framework suggests that people vary in five psychological dimensions: extraversion, neuroticism, openness to experiences, agreeableness, and conscientiousness (Ehrenberg et al., 2008).

Extraversion is a socially-oriented set of personality characteristics related to general needs for belongingness and is associated with talkativeness and adventurousness. Neuroticism, on the other hand, reflects one's capability to control and regulate oneself emotionally; high neuroticism tends to be related to depression and pessimism. Openness to experiences pertains to the need for novelty and change and is typically associated with the impressions of being adventurous. Agreeableness refers to the extent to which one would defer to others in social/conflictual circumstances and is often seen as an essential ingredient for friendship and trust. Lastly, conscientiousness is related to a desirable work ethic and an attitude to strive for achievement; hence, it is often sought after in organizational and leadership settings. These five dimensions of personality are widely accepted as the foundation of understanding an individual's behavior and personality characteristics.

Equipped with this set of criteria about individual differences, researchers surveyed people from over 20 countries about their Big Five personality traits and correlated those with their general frequency of social media use, social media use for news, and social media use for social interaction (Gil de Zúñiga et al., 2017). The results revealed some universal patterns: extraversion, agreeableness, neuroticism, and conscientiousness are positively related to all three types of social media use. But more importantly, across all countries, people with high extroversion tend to be more likely to use social media for informational and social purposes, and high agreeableness is consistently connected with social media use for news. Taking one step further, one might utilize information about people's personality traits to predict their specific social media platform use. For example, research found that extraverts also tend to be highly engaged users on TikTok (Meng & Leung, 2021), and people who are high in extraversion, conscientiousness, and openness to experience are more likely to use Tinder than others (Timmermans & De Caluwé, 2017).

As one might have expected, because of the high predictive power of users' psychological traits and their social media use, many industries are keen to obtain these types of data about their users for marketing, advertising, and various other purposes. For example, in the infamous case of Cambridge Analytics, the British social media analytic company was found to be actively involved in harvesting Facebook users' data, whereby they could predict the susceptible population to distribute targeted political messages during the 2016 US presidential election. No wonder WIRED magazine exclaimed that "data is the new oil of the digital economy" (Toonders, 2014). We will explore more of this point in Part II and Part III of the book.

> **Combating FoMO**
>
> One common psychological experience that has been found among compulsive social media users is Fear of Missing Out (FoMO). Psychologically, FoMO can be seen as deficiencies stemming from unfulfilled intrinsic needs such as relatedness (the desire to connect with others), competence (the need to accomplish personal goals), and autonomy (the need to think freely and take actions) (Przybylski et al., 2013). In the context of social media, one might experience FoMO in numerous settings, such as seeing photos of friends hanging out, not knowing what is going on among others, or knowing about others' achievements. Studies have shown that social-media-induced FoMO is tightly related to negative emotions such as anxiety, irritation, jealousy, and is likely leading to low self-esteem, loneliness, depression, lack of sleep, and compulsive social media usage (Tandon et al., 2021).
>
> So, what can we do about it? For starters, there is some evidence showing that limiting social media use to 10 minutes, per platform, per day would significantly reduce FoMO over a three-week time span (Hunter et al., 2018). Others have developed more comprehensive approaches that involve some technical elements such as autoreply, filtering, status, and education (The FoMO-R Method: www.ncbi.nlm.nih.gov/pmc/articles/PMC7504117/bin/ijerph-17-06128-s001.pdf). Finally, clinical psychologists have also proposed a mindfulness-based intervention that focuses on training users to be more aware of their attention. The steps involve secular mindfulness meditation practices, cognitive reflections on social media use, and monitoring of one's attitude during social media use (see Weaver & Swank, 2019, in the reference for a detailed description of the technique).

Lurkers versus posters

Aside from the fundamental psychological traits of users, we might also see social media users through the lens of their online behaviors. In particular, the dichotomy of lurkers versus posters is uniquely useful. Lurkers refers to the kinds of users who are silently online or passively viewing content for most of their time spent on a platform; whereas poster refers to users who are actively engaged in online interactions and content posting.

You would be absolutely correct in expecting posters to be the user group disproportionally favored by industry practitioners and platform owners. For one, posters generate an enormous amount of data online, which, again, are the bread and butter for social media marketing and advertising. Also, posters generate online traffic, reflecting the popularity of a platform. Executives of social media companies are trained to focus on a long list of "user engagement" metrics, such as likes, comments, shares, and posts. And posters are the key contributors to engagement on virtually all social media platforms. Lastly, some scholars argue that posters are beneficial for democracy, as their engagement in online discussions and debates about policies and politics can lead to higher voter turnout and increased awareness of critical social issues.

If posters are generally welcomed almost across the board, why would some social media users choose to be lurkers? In fact, lurkers are the *de facto* majority among all users: in the business sector, there exists the so-called "1–9–90 rule," which implies that in any given online community, 90% of the users consume content passively, 9% of users contribute to content editing, and merely 1% of users actively create new content (Reed, 2020). And research studying the actual posting behaviors of people on Twitter, for instance, confirmed that the vast majority of users simply browse content without posting anything (Gong et al., 2015).

Lurkers lurk for four main reasons (Sun et al., 2014). First, users lurk because of the environment of the online community. As we will discuss in the next chapter, platforms are not equally designed: they may vary in terms of the quality of information, user interactivity, and response rate, among others, all of which affect users' willingness to participate. Secondly, people's personal characteristics play a role in determining their social media activities. In general, users who are less confident about their ability to produce content and interact with others (i.e., high self-efficacy) are more likely to become lurkers. Or, if one's motivation on a platform is simply to search for information, then being a lurker may be sufficient for his or her need. The third reason for lurking has relevance to relational considerations. Users who perceive low intimacy, a lack of commitment to the group, or simply feel not welcomed on a platform may not feel the necessity of actively participating in it. (This is also why nearly all social media platforms claim that they are devoted to building a welcoming community for all users.) Finally, some users may choose to lurk for privacy and safety concerns. Those who are more sensitive about sharing private information online may inhibit their social media activities. Conversely, those who are more concerned, or in some cases, more educated about online safety issues, will lurk more than their counterparts.

> **Exercise**
>
> Assuming you are the owner of a major social media platform, what strategies would you consider implementing to facilitate users posting new content, given the importance of posters to your platform? Does the type of social media platform matter when considering these strategies? (As a refresher, see Chapter 1 for typologies of social media.) Try searching online news and specific social media companies' announcements for real-world examples. Once you have a list of strategies, perform a quick cost-benefit analysis. Are there any strategies that are relatively easier to implement than others? Which strategies might be most effective?

Trolling and cyberbullying

As desirable as posters can be for maintaining platforms' traffic, unchecked posts can sometimes cause great trouble and harm to platform users. In that regard, trolling and cyberbullying are two of the most destructive posting behaviors in the cybersphere.

Trolling, by definition, refers to "a deviant, malicious or antisocial online behavior with motives to disrupt conversations and trigger conflict" (Gylfason et al., 2021, p. 1). In practice, trolling is manifested by deception, aggression, and disruptive online activities that aim to hurt, shock, and offend targets. More often than not, trollers tend to perceive a sense of accomplishment when their behaviors get attention from others. As a pathological online behavior, trolling tends to be found among those with "Dark Tetrad personality"; namely, Machiavellianism (the intention to control and manipulate others), narcissism (feeling that one is better than others), and sadism (enjoying the suffering of others). Unfortunately, trolling is not uncommon on social media; it happens frequently around online debates on contentious social and political issues and is often related to hate speech, racism, misogyny, homophobia, and the like (Hannan, 2018).

The other behavior closely related to trolling is cyberbullying, or cyberharassment. Cyberbullying can be defined as "willful and repeated harm inflicted through the use of computers, cell phones, and other electronic devices" (Hinduja & Patchin, 2014). Symptomatically, it entails a range of malicious online actions such as flaming, outing, trickery, sexting, and cyberstalking, among many others (Kowalski et al., 2014). However, unlike trolling, which tends to happen to users just episodically, cyberbullying refers to harmful actions of a repeated nature. Additionally, despite happening among users of all ages, cyberbullying happens mostly between teens and adolescents and often results in severe consequences within this population. And victims of cyberbullying often report depression, suicidal ideation, and various negative emotions.

In recent years, cyberbullying and trolling have grown to be prominent social issues both in the US and worldwide, partially due to the ubiquity of social media and the anonymity offered by many of those services. As more and more policymakers, academics, and average users are awaking of these problems, demands for social media companies to add more strict regulations and to take actions to protect their users are rapidly growing. To their credit, some platforms are taking action. For instance, to protect teens from predatory direct messages and harassment, TikTok disabled the direct messaging feature for all users under 16 (Collins, 2020); and more recently, the platform also automatically switched off direct messaging for new users aged between 16 to 17. However, these actions alone will not be sufficient; as trolling and cyberbullying evolve, social media companies will need to install much more sophisticated guardrails to fend off ill-intentioned users and sometimes organized entities.

The iGen

While we are on the subject of cyberbullying and teens, let's zero in on teens' use of social media. Aside from polling the US population about their social media use (as we have encountered at the beginning of the chapter), the Pew Research Center also conducts periodic surveys regarding US teens' (ages 13 to 17) use of social media. And here are some of the highlights from their 2022 survey (Vogels et al., 2022):

- The most popular social media platforms among teens are YouTube (95%), TikTok (65%), Instagram (62%), and Snapchat (59%). The percentage of teens who use Facebook declined dramatically from 71% in 2014 to 32% in 2022.
- Nearly all teens (97%) in the US say they are on the Internet "almost constantly."
- Nearly all teens (95%) in the US have access to a smartphone in 2022.
- 54% of teens indicated that it would be hard for them to give up social media.
- Teen girls are more likely than teen boys to be on TikTok, Instagram, and Snapchat, while boys are more likely to be on Twitch, Reddit, and YouTube.

One thing that appears to be clear from these survey results is that today, a generation of US teens are growing up surrounded by digital technology—they are the digital natives. These teens have a unique

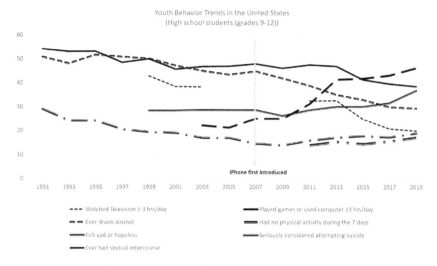

Figure 2.2 Youth behavior trends in the United States

(*Data Source*: CDC The Youth Risk Behavior Surveillance System)

perspective on the digital world, and understanding their experiences can give us valuable insights into the areas that require more attention in the future.

In her 2017 book *iGen*, psychologist Jean Marie Twenge presents a compelling case for the ways in which technology—specifically, smartphones—have shaped the behavior and experiences of teens born between 1995 and 2012. More specifically, Twenge's research reveals a decline in traditional markers of adolescent development such as dating, sexual activity, and driver's license ownership. Alongside this, a rise in feelings of loneliness, sleep deprivation, and disinterest in social interactions at work is observed (see Figure 2.2). The cause of all these concerning trends, according to Twenge's diagnosis, is the introduction of smartphones, particularly the iPhone in 2007.

Exercise

Use this 15-item quiz below from Twenge (2017, p. 11) to see how "iGen" you are.

You should answer each question with either "yes" or "no."

1. In the past 24 hours, did you spend at least an hour total texting on a cell phone?
2. Do you have a Snapchat account?

3. Do you consider yourself a religious person?
4. Did you get your driver's license by the time you turned 17?
5. Do you think same-sex marriage should be legal?
6. Did you ever drink alcohol (more than a few sips) by the time you turned 16?
7. Did you fight with your parents a lot when you were a teen?
8. Were more than 1/3 of the other students at your high school a different race than you?
9. When you were in high school, did you spend nearly every weekend night out with your friends?
10. Did you have a job during the school year when you were in high school?
11. Do you agree that safe spaces and trigger warnings are good ideas and that efforts should be made to reduce microaggressions?
12. Are you a political independent?
13. Do you support the legalization of marijuana?
14. Is having sex without much emotion involved desirable?
15. When you were in high school, did you feel left out and lonely fairly often?

Scoring Note: Give yourself 1 point for answering "yes" to questions 1, 2, 5, 8, 11, 12, 13, 14, and 15. Give yourself 1 point for answering "no" to questions 3, 4, 6, 7, 9, and 10. The higher your score, the more iGen you are in your behaviors, attitudes, and beliefs.

The iPhone effect

Perhaps you are not willing to go so far as to pin down smartphones as the actual cause of these changes among teens just yet. But the empirical evidence linking smartphone use and certain troubling consequences is more robust than most people would like to admit. Foremost among various considerations, many would agree that the presence of mobile devices at times does harm our in-person social interactions. For example, in a study titled "The iPhone Effect," Misra and colleagues (2016) trained research assistants to observe 200 participants in coffee shops and cafes in the Washington D.C. area as they held casual conversations. The assistants recorded whether either participant had a mobile device on the table or in their hand during the conversation. After the conversation, participants completed a questionnaire rating their experience. It was discovered that conversations with mobile devices present were rated as less fulfilling compared to those without mobile devices. Additionally, individuals with close relationships to their conversation partners were more affected by the presence of mobile devices.

If the mere presence of a phone in interpersonal conversations could impede the way we interact with each other, are we better off leaving our phones behind? To tackle this question, researchers invited 40 iPhone users into a lab to perform word search puzzles while wearing a wireless blood pressure monitor cuff (Clayton et al., 2015). However, as the experiment progressed, researchers used an excuse to take some participants' phone away and placed them at a 4-foot distance. And, as the participants proceeded with their tasks, researchers actually called these subjects on their phones. (Participants had already provided the number to researchers ahead of time.) Somewhat expectedly, participants who could not access their ringing iPhones during the word search puzzle showed increased heart rate and blood pressure, accompanied with self-reported feelings of anxiety and unpleasantness, and decreased cognition. What about those with their phones in possession all along? Their heart rate and blood pressure levels returned to baseline after a temporary increase, coupled with an increase in cognitive performance. So, leaving your phones behind will not be the solution, at least not during solving word search puzzles.

Smartphones don't just affect teens' behaviors and ways of thinking but also their emotions. For the prolonged use of smartphones and social media tend to increase feelings of comparison and pressure to conform to certain social norms, two psychological conditions that are uniquely salient among teens. And, depending on the context in which teens experience these feelings, the negative impact on teens' emotional well-being seems unavoidable. In fact, as we delve deeper into the psychological effects of smartphone use, the evidence is becoming increasingly clear. A comprehensive review of 23 studies on the link between problematic smartphone use and mental health issues revealed a consistent correlation between excessive smartphone usage and feelings of anxiety and stress (Elhai et al., 2017). In other words, the more an individual struggles with problematic phone habits, the higher their levels of anxiety and stress tend to be.

Motivations of social media use

But you would argue that certainly not all types of smartphone use are problematic. Considering all the applications, particularly social media apps on our phones, could it be that people use social media differently, and if so, why do people use media, and what do they use them for? These are the central questions to which the Uses and Gratifications (U&G) research seeks to answer.

As an influential field in media effect research, U&G approaches media use in terms of the gratification or psychological needs of the individual (Blumler & Katz, 1974). It assumes that audiences consciously choose the medium that could fulfill their needs and that they are able to recognize their reasons for making media choices. In the context of social media, individual users will continue to be engaged with social networking sites if their needs are fulfilled by specific platform usage. For example, in an early study of college-aged Facebook group users, researchers found that people use the platform for information acquisition about campus/community, entertainment/recreation, social interaction with friends and family, and peer pressure/self-satisfaction (Park et al., 2009). Snapchat, in contrast, serves users in attaining emotional support from others, looking for advice on important decisions, seeking help to solve problems, satisfying the need to socialize, venting negative feelings, and connecting with others (Phua et al., 2017).

Beyond identifying the potential motives of use, the U&G approach can be extremely instructive used in conjunction with social media users' behavioral data, such as the amount of time spent on the platform and how they interact. Recall the Pew Research Center's polling on teens' use of social media earlier, which suggests that teen boys are more likely to be on Twitch than girls? One U&G research surveyed over a thousand Twitch users and correlated five types of motivations (information seeking/learning, enjoyment, personal recognition, companionship and emotion sharing, as well as relaxation) with how people actually use Twitch (Sjöblom & Hamari, 2017). It was revealed that the extent to which using Twitch fulfills the needs for information seeking, relaxation, and companionship predicts one's time spent on Twitch. Furthermore, companionship is critical for users' decision to subscribe to specific streamers, and the information seeking is positively associated with the total number of individual streamers one chooses to watch. Again, if you are a platform owner, insights like this are precious for increasing the platform's overall engagement. As users, though, we might want to reflect upon how our behaviors on social media are shaped by our own desires and needs.

Keywords: digital divide; digital immigrants; Big-Five personality framework; FoMO; posters, lurkers; trolling; cyberbullying; iGen; iPhone effect; Uses & Gratifications.

References

Blumler, J. G., & Katz, E. (1974). The uses of mass communications: Current perspectives on gratifications research. Beverly Hills, California, Sage Publications, 1974, 318 pp. *Public Opinion Quarterly*, *40*(1), 132–133. https://doi.org/10.1086/268277

Clayton, R. B., Leshner, G., & Almond, A. (2015). The extended iSelf: The impact of iPhone separation on cognition, emotion, and physiology. *Journal of computer-mediated communication*, *20*(2), 119–135.

Collins, J. (2020). TikTok introduces family pairing. *TikTok.com*. https://newsroom.tiktok.com/en-us/tiktok-introduces-family-pairing

Dargin, J. S., Fan, C., & Mostafavi, A. (2021). Vulnerable populations and social media use in disasters: Uncovering the digital divide in three major US hurricanes. *International Journal of Disaster Risk Reduction*, 54, 102043. https://doi.org/10.1016/j.ijdrr.2021.102043

Ehrenberg, A., Juckes, S., White, K. M., & Walsh, S. P. (2008). Personality and self-esteem as predictors of young people's technology use. *Cyber Psychology & Behavior*, 11(6), 739–741. https://doi.org/10.1089/cpb.2008.0030

Elhai, J. D., Dvorak, R. D., Levine, J. C., & Hall, B. J. (2017). Problematic smartphone use: A conceptual overview and systematic review of relations with anxiety and depression psychopathology. *Journal of Affective Disorders*, 207, 251–259. https://doi.org/10.1016/j.jad.2016.08.030

Gil de Zúñiga, H., Diehl, T., Huber, B., & Liu, J. (2017). Personality traits and social media use in 20 countries: How personality relates to frequency of social media use, social media news use, and social media use for social interaction. *Cyberpsychology, Behavior, and Social Networking*, 20(9), 540–552. https://doi.org/10.1089/cyber.2017.0295

Gong, W., Lim, E. P., & Zhu, F. (2015). Characterizing silent users in social media communities. *Proceedings of the International AAAI Conference on Web and Social Media*, 9(1), 140–149. https://doi.org/10.1609/icwsm.v9i1.14582

Gylfason, H. F., Sveinsdottir, A. H., Vésteinsdóttir, V., & Sigurvinsdottir, R. (2021). Haters gonna hate, trolls gonna troll: The personality profile of a Facebook troll. *International Journal of Environmental Research and Public Health*, 18(11), 5722. https://doi.org/10.3390/ijerph18115722

Hannan, J. (2018). Trolling ourselves to death? Social media and post-truth politics. *European Journal of Communication*, 33(2), 214–226. https://doi.org/10.1177/0267323118760323

Hinduja, S., & Patchin, J. W. (2014). *Bullying beyond the schoolyard: Preventing and responding to cyberbullying*. Corwin Press.

Hunter, J. (2018). The development of a single item FOMO (fear of missing out) scale. *Current Psychology*, 39(4), 1215–1220. https://doi.org/10.1007/s12144-018-9824-8

Kowalski, R. M., Giumetti, G. W., Schroeder, A. N., & Lattanner, M. R. (2014). Bullying in the digital age: A critical review and meta-analysis of cyberbullying research among youth. *Psychological Bulletin*, 140, 1073–1137. https://doi.org/10.1037/a0035618

Meng, K. S., & Leung, L. (2021). Factors influencing TikTok engagement behaviors in China: An examination of gratifications sought, narcissism, and the big five personality traits. *Telecommunications Policy*, 45(7), 102172. https://doi.org/10.1016/j.telpol.2021.102172

Misra, S., Cheng, L., Genevie, J., & Yuan, M. (2016). The iPhone effect: The quality of in-person social interactions in the presence of mobile devices. *Environment and Behavior*, 48(2), 275–298. https://doi.org/10.1177/0013916514539755

Park, N., Kee, K. F., & Valenzuela, S. (2009). Being immersed in social networking environment: Facebook groups, uses and gratifications, and social outcomes. *Cyberpsychology & Behavior*, 12(6), 729–733. https://doi.org/10.1089/cpb.2009.0003

Phua, J., Jin, S. V., & Kim, J. J. (2017). Uses and gratifications of social networking sites for bridging and bonding social capital: A comparison of Facebook, Twitter, Instagram, and Snapchat. *Computers in Human Behavior*, 72, 115–122. https://doi.org/10.1016/j.chb.2017.02.041

Prensky, M. (2001). Digital natives, digital immigrants part 1. *On the Horizon*, 9(5), 1–6. https://doi.org/10.1108/10748120110424816

Przybylski, A. K., Murayama, K., DeHaan, C. R., & Gladwell, V. (2013). Motivational, emotional, and behavioral correlates of fear of missing out. *Computers in Human Behavior*, 29(4), 1841–1848. https://doi.org/10.1016/j.chb.2013.02.014

Reed, C. (2020). How to leverage the 1–9–90 rule and become a leader on LinkedIn. *Forbes.com*. www.forbes.com/sites/forbesbusinesscouncil/2020/07/10/how-to-leverage-the-1-9-90-rule-and-become-a-leader-on-linkedin/?sh=69f6f6eb7d32

Rogan, J. (2022, August 25). Mark Zuckerberg (No. 1863). In *The Joe Rogan experience*. https://open.spotify.com/show/4rOoJ6Egrf8K2IrywzwOMk?si=a436722faaf748f5

Sjöblom, M., & Hamari, J. (2017). Why do people watch others play video games? An empirical study on the motivations of Twitch users. *Computers in Human Behavior*, 75, 985–996. https://doi.org/10.1016/j.chb.2016.10.019

Sun, N., Rau, P. P. L., & Ma, L. (2014). Understanding lurkers in online communities: A literature review. *Computers in Human Behavior*, 38, 110–117. https://doi.org/10.1016/j.chb.2014.05.022

Tandon, A., Dhir, A., Almugren, I., AlNemer, G. N., & Mäntymäki, M. (2021). Fear of missing out (FoMO) among social media users: A systematic literature review, synthesis and framework for future research. *Internet Research*. https://doi.org/10.1108/intr-11-2019-0455

Timmermans, E., & De Caluwé, E. (2017). To Tinder or not to Tinder, that's the question: An individual differences perspective to Tinder use and motives. *Personality and Individual Differences*, 110, 74–79. https://doi.org/10.1016/j.paid.2017.01.026

Toonders, J. (2014). Data is the new oil of the digital economy. *Wired*. www.wired.com/insights/2014/07/data-new-oil-digital-economy/

Twenge, J. M. (2017). *iGen: Why today's super-connected kids are growing up less rebellious, more tolerant, less happy – and completely unprepared for adulthood – and what that means for the rest of us*. Simon and Schuster.

Van Deursen, A. J., & Van Dijk, J. A. (2014). The digital divide shifts to differences in usage. *New Media & Society*, 16(3), 507–526. https://doi.org/10.1177/1461444813487959

Vogels, E. A., Gelles-Watnick, R., & Massarat, N. (2022). Teens, social media and technology 2022. *Pew Research Center*. www.pewresearch.org/internet/2022/08/10/teens-social-media-and-technology-2022/

Weaver, J. L., & Swank, J. M. (2019). Mindful connections: A mindfulness-based intervention for adolescent social media users. *Journal of Child and Adolescent Counseling*, 5(2), 103–112. https://doi.org/10.1080/23727810.2019.1586419

Whittaker, A., Densley, J., & Moser, K. S. (2020). No two gangs are alike: The digital divide in street gangs' differential adaptations to social media. *Computers in Human Behavior*, 110, 106403. https://doi.org/10.1016/j.chb.2020.106403

3 Social media design and affordance

On May 26, 2021, the social media giant Meta (parenting company of popular applications such as Facebook, Instagram, and WhatsApp) announced that they would allow users to hide "like" counts on their posts and others' posts that appear on users' feeds. To many tech industry observers, this was not a surprise. For years, both Facebook and Instagram have been facing an ever-growing onslaught of criticisms about how their platforms fuel a pressurized and toxic online environment that causes mental health issues and addictive use. And allowing users to hide the undesirable like counts, as many pundits argued, shows the company's willingness to engage the public concern through better design and features of their products.

Of course, social media design decisions are not always reactive in nature, at least not from the founders' perspectives. In an interview at South by Southwest 2019, Instagram founders Kevin Systrom and Mike Krieger shared their designing philosophies behind the original Instagram, which was to help users bring their followers to their experiences. (Both founders left the company in 2018.) During that interview, Systrom and Krieger stressed that they expected users to engage in the app in the most authentic ways; hence, the limited photo editing functionalities and the more vivid story feature, which presumably allow people to post more freely and casually without the fear of being judged.

Exercise

Most tech companies nowadays have a senior position commonly known as Chief Design Officer (CDO), Product Design Director, or simply Head of Product. People who hold these positions typically work closely with the companies' CEOs to envision and implement the design principles for their products. Now, pick a social media platform that you are familiar with or interested in and investigate who the person currently assuming the role of CDO is. Search online for this person's public talks or interviews, where he or she might have mentioned details about the design philosophy for the platform. Given your experiences with this particular social media platform and what you have learned about the company's design principles, do you think that the company's products (i.e., features and functionalities) live up to what was said about them in public? Share your investigations with the class.

In our materialistic world, design is everywhere—from the chair you are sitting in to the keyboard you are typing on. A poorly designed product is often noticeable or palpable, as in the case of a nonmalleable office chair or the infamous butterfly keyboard. But to some, the claim that a virtual product (such as a piece of software or a digital feature) that also involves design may not appear to be so intuitive; after all, virtual products are often devoid of users' bodily experiences, obfuscating the distinctive experiences individuals might have over a long period. So, how are designs being realized on social media platforms? What is the rationale for social network sites when adopting similar functionalities such as the like button, commenting, and sharing features, self-destructing messages, or more recently, short videos (e.g., Reels, YouTube Shorts)? In this chapter, we will

Figure 3.1 A new feature of Instagram allows hiding "like" counts

36 Part I

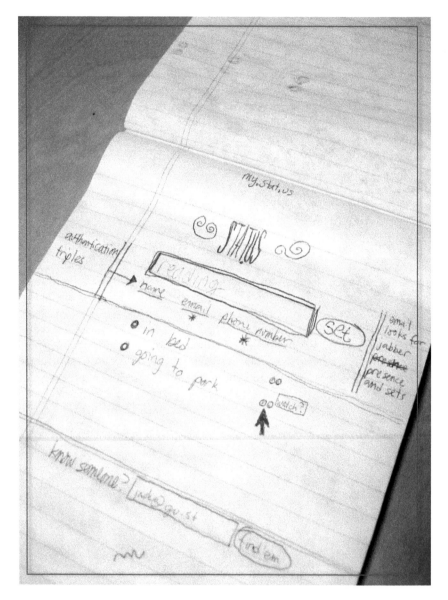

Figure 3.2 Jack Dorsey's first sketch of Twitter
(*Source*: Jack Dorsey/Flickr)

explore what design means in the realm of social media. In this process, we will also contemplate in what context and condition features of social media can bring in desirable outcomes. As we move to the latter part of the chapter, we will illustrate how issues of social media design manifested in a specific area which we can all speak of; namely, work.

Affordance

One idea that is crucial to our understanding of social media design is the notion of affordance, a concept first proposed by psychologist James Gibson in 1979. In his seminal book *The Ecological Approach to Visual Perception*, Gibson wrote, "the *affordances* of the environment are what it *offers* the animal, what it *provides* or *furnishes*, either for good or will" (2015, p. 119; emphases are original). Writing as a researcher of human/animal visual perception, the core argument of his book was that, as humans, we do not perceive things and our physical surroundings as such, but rather, perceive them through their affordances and the potential for action they may provide to us. To illustrate, think of rivers or lakes; to our common ancestors, these water bodies imply many things: they could be sources for drinking, places for play, means of transportation, or perhaps, an early version of air conditioning. As such, the water indicates myriad possible actions. By the same token, a tennis ball can be used for sports, as an alternative pet toy, as a dryer ball for laundry, and much more.

This idea of affordance was later polished and populated by cognitive scientist and industrial design specialist Donald Arthur Norman through the book *The Design of Everyday Things* (1988). In particular, Donald Norman (1988) distilled the notion of "perceived affordance" to suggest that designers can and should "indicate how the user is to interact with the device" (p. 9). You can see how influential this concept became in software and app design. Take our phones, for instance. On any given day, people receive hundreds, if not thousands, of direct messages, emails, texts, likes, and comments from various apps in their phones. How do users manage to figure out which app to open up to check for those incoming messages? Well, we do not think about it at all because the red dots tell everything. Known as "badges" in the industry, these little red dots are commonly seen at the corners of the app icons (often with a number), reminding us that there are new messages from the noted app that demand our attention. In many ways, the red dot is one of the most ingenious designs in the software industry. As *The New York Times* staff writer John Herrman (2018) wrote:

> Dots are deceptively, insidiously simple: They are either there or they're not; they contain a number, and that number has a value. But they imbue whatever they touch with a spirit of urgency, reminding us that behind each otherwise static icon is unfinished business. They don't so much inform us or guide us as correct us: You're looking there, but you should be looking here. They're a lawn that must be mowed. Boils that must be lanced, or at least scabs that itch to be picked. They're Bubble Wrap laid over your entire digital existence.
>
> (para. 9)

38   Part I

Affordances of social media: the user's perspective

The affordance perspective suggests that the features of social media can be understood through a few key affordances or characteristics that enable certain actions. These affordances can vary depending on the perspective of the user and the context in which social media is being used. Kietzmann and colleagues (2011) identified seven affordances of social media: identity, conversations, sharing, presence, relationships, reputation, and groups (see Figure 3.3). Below, we will discuss these affordances and provide examples to illustrate their meanings.

First, the *identity* affordance refers to the level of flexibility that users have to reveal their identities within a social media platform. Users' identity information can range from basic demographic information, such as age, gender, and occupation, to more sophisticated social identities, such as sexual orientation and political ideology. Some platforms actively encourage

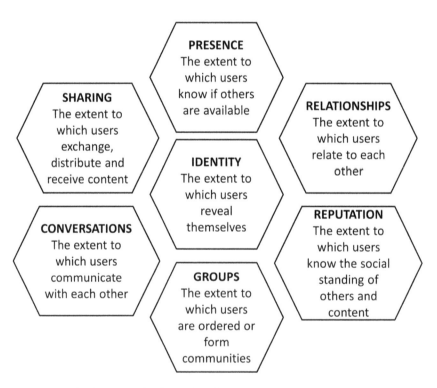

Figure 3.3 The honeycomb of social media

(*Source*: Reproduced with permission from Kietzmann et al., 2011)

or even mandate demographic information (e.g., Facebook and Tinder), while others grant more room for anonymity with the hope that users will be more open to talking about topics of common interest (e.g., Reddit).

The *conversation* affordance suggests the ease with which users can communicate with others on the same platform. For example, Snapchat and WhatsApp are particularly useful for one-on-one or group-based communication, whereas Newsletter or Substack are more suitable for one-to-many broadcasts.

The *sharing* affordance represents the extent to which users can disseminate and exchange information. For example, some contend that early versions of Twitter limited a tweet to 140 characters, which hindered information sharing among users.

The *presence* affordance denotes the degree to which users can tell whether or not other users are accessible. Features such as a typing indicator in messaging apps or an online/offline status icon are examples of high presence-affordance.

The *relationship* affordance suggests the extent to which users can be related to one another and facilitate connections. An example of this is LinkedIn, which allows users to see their organizational affiliations and through whom they might be connected from a targeted member, which potentially facilitates professional relationship building.

The *reputation* affordance denotes the potential for users to identify the standing or "performance" of themselves and others in a social media platform. This can be done through generic metrics such as the number of followers, likes, and comments, or the more narrowly defined product/service reviews and ratings.

Finally, the *group* affordance represents the possibilities for users to form communities and subcommunities. Some platforms allow the assembling of groups that are open and free for users to join and leave (e.g., subreddits and public Facebook groups), whereas others allow users to define group membership or dictate the rules of operation for a given group (e.g., private discord groups or clubhouse meetings).

The affordances of social media can serve a plethora of useful purposes for a diverse array of individuals, organizations, and society at large. Discerning the affordances of a social media platform can help users make informed decisions about which platforms to use and how to use them optimally. For businesses, it can provide insights into the most appropriate platforms to reach their target audience and effectively promote their products or services. Moreover, comprehending the affordances of social media can be instrumental for researchers, educators, and policymakers, empowering them to better comprehend the profound impact of social media on society and how to counteract any adverse effects. To summarize, understanding the affordances of social media can enable individuals and organizations to make sagacious choices about how to use these platforms effectively, for their own advantage and that of society.

> **Exercise**
>
> We've just laid out a framework through which one can make parallel comparisons across social media platforms. Now try to make use of this framework for yourself. Specifically, take two social media platforms you are familiar with and assess their perceived level of affordance respectively. For each affordance, rate the platform on one of the three levels: low, medium, and high. Once you complete this assessment, contemplate the following question: Is it possible to introduce one or two features (real or imagined) into one platform, such that it will be made to be more like the other platform in terms of its palette of affordances?
>
> Consider one specific example: the recent boom of short-video platforms such as TikTok prompt many legacy social media companies to launch similar features. Among a list of followers, Instagram's installment of the "reels" feature (which enables the creation and consumption of short, multi-clip videos, as well as algorithmically recommended videos) was rather turbulent. In the early days of its induction, high-profile celebrities such as Kylie Jenner and Kim Kardashian openly denounced reels, citing that such a feature makes Instagram losing its own identities (Vanian, 2022). What do you make of this criticism? Was it a bad move on Instagram? To what extent do the reels feature to change the dynamics of Instagram among its average users? And finally, what affordances were altered with the induction of this one feature on Instagram?

The influence of platform affordances on social interactions

If the implication about the affordances perspective is still a little too vague, let us bring the concept closer to home. For average social media users like you and me, authenticity is of the most crucial value in our daily online interactions. After all, most people dislike the feeling of second-guessing the content they post. Similarly, seeing others talking about themselves spuriously just makes the whole process of building real connections online difficult. So, is it really possible, as the founders of Instagram argued, to engineer a sense of authenticity in social media platforms?

A group of German scholars recently put Instagram's design principle on trial. More specifically, Kreling and colleagues (2022) recruited 202 Instagram users online and instructed them to review their recent content shared via Instagram's *Stories* and *Post* feature. Participants rated each piece of content on its perceived authenticity and spontaneity level. Results indicated that content transmitted via the stories feature, on average, generated a higher level of perceived spontaneity than those shared via post, which contributed to a heightened sense of authenticity. This suggests that the design and implementation of certain new features can indeed change user experiences on the platform.

Nevertheless, as alluded to earlier, users' experience on social media platforms is an accumulative response to a multitude of affordances rather than a single feature. For example, to understand the popularity of TikTok and its users' "just be you" mentality, Barta and Andalibi (2021) arranged a set of in-depth interviews with 15 TikTok users. They found that the normative authenticity that permeates TikTok users comes from a combination of the platform's anonymity, content association, and the video modality: users reported fewer concerns in expressing themselves due to the platform's low requirement for personal information. Meanwhile, the algorithmically associated content allows users to discover and respond to content of close interests, reducing users' perceived risks about content sharing and emotional expression. Finally, users indicated the richness of TikTok's video mode allows straightforward depiction of raw emotion and actions, which is also an essential ingredient of authenticity. This may explain why Instagram's attempt to catch up with TikTok by introducing full-screen videos and photos was deemed unsatisfactory in the eyes of many Instagram users (Clark & Perrett, 2022).

But a platform's affordances don't affect all of its users evenly. For example, Yeshua-Katz and Hård af Segerstad (2020) studied online support groups for marginalized communities and found that the affordances of group membership (i.e., open accessed vs. closed groups) and identity (anonymous vs. non-anonymous) could either improve visibility and attract new members or create a "digital haven" for existing members. However, these same affordances could also limit the group's ability to educate the public and influence those with opposing views when the group became closed off. This demonstrates how the affordances of a platform can be both helpful and hindering for its users at the same time.

Finally, we should not forget that affordance is a non-static notion. As platforms innovate and learn from each other, new forms of affordances will be ushered in which can potentially disrupt or transform the current dynamics among its users. One such example is the emergence of the so-called transactional affordance, defined as "how the technical features enabling an economic exchange are realized through the contextual awareness and opportunities for specific types of action afforded to individual users" (Manzerolle & Daubs, 2021, p. 1281). Materialistically, transactional affordance is manifested by features related to payment, transaction, and purchase (think, for example, Instagram's Shop, Snapchat's Scan AR, and YouTube's Super Chat). And through these features, transactional affordance becomes a substantial enabler to the current social-media-influencer economy, in which influencers with large or small fan bases find ways to monetize and grow their online traction or brands. Needless to say, successful influencers tend to be more proficient at leveraging this and various other affordances of a platform to engineer a sense of "performative authenticity" (Shtern et al., 2019), so as to make themselves appear trustworthy and relatable among their followers. Chapter 11 will explore the topic of influencers and their relation to the notion of authenticity in greater detail.

Affordances of social media: the organizational perspective

The framework of social media affordance is not only instrumental for our understanding of social media in most social settings but can also shed light on the way we work and collaborate with others. During the height of the COVID-19 pandemic, for instance, many business organizations across the globe quickly adapted to the lockdown and quarantine restrictions by allowing their employees to work from home. One condition that made this transition possible is the availability of a slew of newly launched social media platforms designed specifically for employee communication, such as Zoom, Slack, Microsoft Teams, and ByteDance Lark. One might ask, why these platforms? After all, enterprise social media have been around for decades (some of the established "old" ones include, for instance, Salesforces' Chatter, IBM's Connections, and Cisco's Webex Social). What makes these newcomers to stand out? And perhaps more pertinent to our interests, in the modern-day knowledge-based economy, what roles are social media platforms serving in organizations? Is it always beneficial to rely on social media for employee collaborations? If not, what are the pitfalls that organizations should be aware of while introducing a social media platform to their employees?

Does working from home make sense?

Before the pandemic forced the idea of working from home (WFH) to become a mundane necessity, most businesses were not so keen about leaving their employees working in an unchecked private environment. The assumption was that, if people get to work in a comfortable (and perhaps distractive) home environment, slacking would be unavoidable. But is that the case?

In a highly publicized research, Sandford economists collaborated with a NASDAQ-listed travel agency located in Shanghai with over 16,000 employees to test the real economic consequences of WFH. Specifically, the company's call center employees were contacted about the possibilities for WFH. Among nearly 500 employees who expressed interest, half were randomly selected for WFH, and the rest worked in the office for 9 months. Contrary to the common wisdom, the results showed that employees who chose WFH had, on average, roughly a 13% increase in work performance and a 50% drop in employee quit-rates. The overwhelming evidence also led the company to roll out the option to WFH to the whole firm.

Nevertheless, the decision to allow employees to work remotely involves more than just financial considerations. The availability of information communication technology is also a key factor. In a study of 363 employees who worked remotely during the COVID-19 pandemic, researchers found that anxiety about work-related communication technology negatively impacted job performance, while non-work-related technology such as smartphones and social media were distracting and reduced productivity (Prodanova & Kocarev, 2021). These findings suggest that, if organizations want to transition to remote work successfully, they should provide proper training and preparation to ensure that productivity is not negatively affected.

Affordances of social media in the workplace

In an attempt to address the earlier questions, Majchrzak and colleagues (2013) proposed an affordance-based perspective to illustrate how social media foster publicly visible knowledge conversations in the workplace. Specifically, they laid out the ground for four affordances of social media in the work setting: metavoicing, triggered attending, network-informed associating, and generative role-taking (see Figure 3.4).

First, metavoicing refers to the ability for employees to engage in ongoing online discussions through various ways of reactions. And the target to which one chooses to react can be numerous, such as others' presence, profile, content, and activities. For example, on Zoom, participants can vote on ideas, share a document via screensharing, and comment on ideas through chats. All of these reactions can be particularly beneficial in the process of knowledge creation. As employees' involvements accumulate, other forms of participation would follow suit. Similarly, a high level of

Affordance	Sample features (platform examples)	Organizational benefit	Organizational pitfall
Meta-voicing	Polls, Chat (Zoom)	Drive participation and identifying areas of problem/interest	Group think or biased views
Triggered Attending	Notifications (Slack)	Lower effort required to remain engaged	Missing other undefined events or content; Managerial surveillance and coercion
Network-informed associating	Organization chart, file search (MS Teams)	Make discovering and finding connections easy	Attentions may be devoted to the very few social hubs
Generative role-taking	Wiki, groups (Lark)	Allow various ways of organizing	Memory loss and shift of social norms

Figure 3.4 Four affordances of social media in the work settings

participation through metavoicing also helps an organization to quickly identify issues/topics of common interest or significance, which helps to steer the conversation in the right direction. However, a significant issue with metavoicing is that employees might suppress their ideas if they notice that they are of an opinion minority. Without enough people expressing different opinions, unwise groupthink may take root.

The second social media affordance in workplace conversation is triggered attending. As the name suggests, triggered attending entails the ability for members of an organization to stay uninvolved in a given task or conversation until a timely alert or notification occurs. For instance, the notification feature in Slack allows users to get informed when someone replies to a thread that one has been following or uses certain keywords one is interested in. Features like this can significantly reduce the effort for an employee to remain engaged during lengthy conversations. However, one side-effect of such an affordance is that one can only get notified on pre-defined terms, which means that one can still miss important information or event. Moreover, triggered attending may also open doors for some abusive leaders to conduct micromanagement and attract unwanted employee scrutiny, which in the short term, cause employees to disengage, and in the long run, erodes trust in organizations.

> **Enterprise social media for who?**
>
> Is it possible that an enterprise social media designed with triggered attending affordance would do more harm than good? Consider DingTalk, an app developed by the Chinese e-commerce giant Alibaba. As a communication platform, DingTalk shares many functional similarities with its US counterpart, Slack, such as group chat, audio-visual calls, and file exchange. But what really sets DingTalk apart from others is that it seems to prioritize the needs of the managers.
>
> Take the message feature, for example. DingTalk allows managers to draw specific employees' attention with a tagging feature, such that the tagged employee will not only be notified in the app but also receive automated phone calls and text messages. Once the employee reads the manager's note, the manager will receive a "read" receipt, forcing workers to respond almost immediately. Moreover, the app also has a built-in clock in/out feature, which resembles a punch card machine in the offline environment. Yet, it also tracks how long employees have connected to the office wireless network and the times they have disconnected due to lunch and other private reasons.
>
> One would argue that triggered attending at this level will result in employees constantly feeling under surveillance and rushing their work. Unfortunately, such problems seem to be only exacerbating in recent days. News reports suggest that, in the wake of the COVID-19 pandemic and employees' need for work from home, many employers have started to use surveillance software to track their employees' activities in private spaces (Alsever, 2021). One major concern is that employees who work in remote settings don't know what exactly is being recorded and whether or not those recorded information would be used against them in critical moments.

The third social media affordance for workplace communication is called network-informed associating. In organizational settings, employees naturally need to know how people are related and how content is connected. Hence, network-informed associating refers to the ability for users to discover relational and content ties in a given platform. For instance, on Microsoft Teams, users who work within a large organization or across multiple channels can identify how they are related to others through the "organization chart," which displays people's organizational connections. Users can also search content based on document authorship, project membership, and individual role in the organization. Understandably, features like these help users work efficiently in identifying each other's areas of expertise and finding the appropriate person to connect with for specific tasks. However, an unintended consequence of network-informed associating is that it can sometimes lead to the "Matthew effect," where individuals already in advantageous positions (or who have reputations) will draw even more attention, resulting in limited learning and knowledge exposure for members of the organization who urgently need such opportunities. In the longer term, this may discourage new ideas and connections from being established elsewhere.

Lastly, many enterprise social media support knowledge production via generative role-taking. Simply put, generative role-taking denotes the ability for members of an organization to voluntarily take on roles that facilitate the completion of tasks or address specific problems. A prime example is Lark, an enterprise collaboration platform developed by ByteDance (parenting company of TikTok). Through Lark, users can collaboratively work on editing companies' wiki pages or documents for specific tasks. In addition, users can create groups and initiate meetings, depending on their needs, while temporarily assuming administrator or group leader roles. In doing so, users can freely experiment with different ways of organizing within a company, which facilitates cooperation. A major issue, however, is that "memory loss" might happen as people constantly enter and exit groups and teams, resulting in repeated visiting of problems that may have been previously discussed or resolved. Moreover, the fluidity of team members might also lead to quick changes of social norms within groups, which might create conditions for disputes.

In short, these four affordances constitute some of the common ways in which major enterprise social media support knowledge conversations in organizations. Depending on how people engage these platform affordances, their works can be either greatly facilitated or stymied. Rather than relying on trials and errors, both organizations and employees can be better served with a deeper understanding of what their communication technologies might mean to their daily work routine.

The influence of platform affordances on employees'
non-work-related matters

Knowledge sharing, collaboration, and outreach are fundamental to the operations of nearly all organizations. But organizations are constituted by employees who, except for working, also have the needs for socialization, leisure, and sometimes, expressing concerns and grievances. In that light, the extent to which social media affordances address those non-work-centered needs is absolutely central to the success of an organization.

Consider organizational socialization, which refers to the "communicative process through which individuals join organizations, develop expectations, and adopt organizational roles" (Lee et al., 2019, p. 2). Many types of social relationships can exist simultaneously within an organization, but those with coworkers are arguably the most fundamental and impactful. In this context, Guo and colleagues (2021) paid attention to a particular group of workers: new/entry-level employees. In their study, researchers interviewed 17 young, office-based professionals from multiple industries and asked about their use of social media for organization socialization. What they found was that social media affordances enabled and, at the same time, constrained employees' assimilation into their company culture at all stages of organizational relationship development. For example, affordances such as "content visibility and persistence" (similar to the metavoicing discussed earlier) and "network association" (an equivalent to the idea of network-informed associating) facilitated new employees to find helpful information and identify the right coworkers to form relationships. However, the visibility and persistence of previous information also made them hyper-alerted to impression management, to the extent that some would actively edit their profiles and messages to avoid revealing personal matters that may reflect negatively on themselves (e.g., party, drinking, etc.).

Of course, there is no reason to presume that enterprise social media affordances only matter in the socialization of early-career employees: employees at other career stages also face peculiar needs and challenges in connecting with coworkers via social media. This is demonstrated in a study of workplace socialization via WeChat, a popular social media platform used by most Chinese for both social and professional purposes. More specifically, Zhu and Miao (2021) were interested in how the "like" feature in WeChat is being used and interpreted by professionals in China, a culture that places high values on interpersonal relationships. Through extensive in-depth interviews with 52 people of various demographic backgrounds, they show that people's decisions to like coworkers' posts bear many meanings, such as paying respect to senior colleagues, maintaining a good rapport, implying a request, or following social etiquette. As such, the ease with which one can like their coworkers' post, at times, can cause confusion and a sense of coarseness. In short, even a feature as simple as the like button might significantly impact employees' social interactions in non-work-related settings.

Another concern about the use of enterprise social media or communication technologies (from the organization's point of view) is that employees might engage in cyberslacking: a range of deviant behaviors such as viewing sports and entertainment videos, visiting news, and shopping sites, playing online games, or engaging stock trading, all during work hours. Needless to say, activities like these are adversarial to employees' productivity, and hence, are often explicitly discouraged by business organizations. But rather than directly telling employees not to do something (or worse, limiting workers' time using Internet and other digital devices), well-designed enterprise social media may act as non-intrusive measures to help reduce cyberslacking. Studies have shown that employees who can socialize with each other through expressing themselves, receiving colleagues' recognition (via likes and comments), and networking with others on enterprise social media are less likely to engage in cyberslacking; presumably because these social interactions contribute to a sense of workplace social bonding (Luqman et al., 2020; Nivedhitha & Manzoor, 2020).

Finally, successful business organizations also need employees who can freely and effectively express their concerns and grievances. In the traditional workplace settings, employees either have to risk their job security (or relationship with other coworkers) to voice their concerns publicly or place their comments anonymously into a suggestions box, hoping they might get to the right organizational leaders. This conundrum might be addressed by using social media channels that allow employees to share their concerns anonymously within a small targeted group, as this would make employees feel much safer and more confident in deciding to voice out (Mao & DeAndrea, 2019).

Keywords: Affordance; generic affordances of social media; authenticity; transactional affordance; enterprise social media; affordances of social media in workplaces; organizational socialization; cyberslacking.

References

Alsever, J. (2021). Your company could be spying on you: Surveillance software use up over 50% since the pandemic started. *Fortune.com*. https://fortune.com/2021/09/01/companies-spying-on-employees-home-surveillnce-remote-work-computer

Barta, K., & Andalibi, N. (2021). Constructing authenticity on TikTok: Social norms and social support on the "Fun" platform. *Proceedings of the ACM on Human-Computer Interaction*, 5(CSCW2), 1–29.

Clark, T., & Perrett, C. (2022). Instagram is killing a big product overhaul after the Kardashians and other influencers vented frustrations about it. *Businessinsider.com*. www.businessinsider.com/instagram-walks-back-full-screen-feed-changes-kardashians-influencer-backlash-2022-7

Gibson, J. J. (2015). *The ecological approach to visual perception* (Classic ed.). Psychology Press. https://doi.org/10.4324/9781315740218

Guo, Y., Lee, S. K., & Kramer, M. W. (2021). "We work in international companies": Affordances of communication media in Chinese employees' organizational socialization. *Communication Studies*, 72(5), 788–806.

Herrman, J. (2018, February 27). How tiny red dots took over your life. *The New York Times*. www.nytimes.com/2018/02/27/magazine/red-dots-badge-phones-notification.html

Kietzmann, J. H., Hermkens, K., McCarthy, I. P., & Silvestre, B. S. (2011). Social media? Get serious! Understanding the functional building blocks of social media. *Business Horizons*, 54(3), 241–251. https://doi.org/10.1016/j.bushor.2011.01.005

Kreling, R., Meier, A., & Reinecke, L. (2022). Feeling authentic on social media: Subjective authenticity across Instagram stories and posts. *Social Media + Society*, 8(1), 20563051221086235. https://doi.org/10.31234/osf.io/jz3wm

Lee, S. K., Kramer, M. W., & Guo, Y. (2019). Social media affordances in entry-level employees' socialization: Employee agency in the management of their professional impressions and vulnerability during early stages of socialization. *New Technology, Work and Employment*, 34(3), 244–261. https://doi.org/10.1111/ntwe.12147

Luqman, A., Masood, A., Shahzad, F., Imran Rasheed, M., & Weng, Q. (2020). Enterprise social media and cyber-slacking: An integrated perspective. *International Journal of Human-Computer Interaction*, 36(15), 1426–1436. https://doi.org/10.1080/10447318.2020.1752475

Majchrzak, A., Faraj, S., Kane, G. C., & Azad, B. (2013). The contradictory influence of social media affordances on online communal knowledge sharing. *Journal of Computer-Mediated Communication*, 19(1), 38–55.

Manzerolle, V., & Daubs, M. (2021). Friction-free authenticity: Mobile social networks and transactional affordances. *Media, Culture & Society*, 43(7), 1279–1296. https://doi.org/10.1177/0163443721999953

Mao, C. M., & DeAndrea, D. C. (2019). How anonymity and visibility affordances influence employees' decisions about voicing workplace concerns. *Management Communication Quarterly*, 33(2), 160–188. https://doi.org/10.1177/0893318918813202

Nivedhitha, K. S., & Manzoor, A. S. (2020). Get employees talking through enterprise social media! Reduce cyberslacking: A moderated mediation model. *Internet Research*. https://doi.org/10.1108/intr-04-2019-0138

Norman, D. (1988). *The design of everyday things*. Basic Books.

Prodanova, J., & Kocarev, L. (2021). Is job performance conditioned by work-from-home demands and resources? *Technology in Society*, 66, 101672. https://doi.org/10.1016/j.techsoc.2021.101672

Shtern, J., Hill, S., & Chan, D. (2019). Social media influence: Performative authenticity and the relational work of audience commodification in the Philippines. *International Journal of Communication*, 13, 20.

Vanian, J. (2022). Sisters Kylie Jenner and Kim Kardashian urge Instagram to stop copying TikTok. *CNBC*. www.cnbc.com/2022/07/25/kylie-jenner-and-kim-kardashian-urge-instagram-to-stop-copying-tiktok.html

Yeshua-Katz, D., & Hård af Segerstad, Y. (2020). Catch 22: The paradox of social media affordances and stigmatized online support groups. *Social Media + Society*, 6(4), 2056305120984476. https://doi.org/10.1177/2056305120984476

Zhu, H., & Miao, W. (2021). Should I click the "like" button for my colleague? Domesticating social media affordance in the workplace. *Journal of Broadcasting & Electronic Media*, 65(5), 741–760. https://doi.org/10.1080/08838151.2021.1991350

4 Social media economy

In 2016, a BuzzFeed article (which was false) claimed that Twitter planned to change the organization of its timeline from simple, reverse-chronological order to one based on algorithmically determined relevance to each user, causing a widespread reaction among users. Consequently, enraged users stormed to Twitter and inundated the platform with the #RIPTwitter hashtag. It is not arduous to comprehend people's immediate aversion to changes in Twitter's fundamental content delivery mechanism. For many, the term "algorithm" conjures up images of inscrutable and enigmatic operations of big data, government, and business. For others, the algorithm deprives users of control over their own feed and timeline. Furthermore, the opacity and confidentiality surrounding the algorithmic systems are not helping to alleviate public anxiety in any constructive manner.

Algorithm

What is an algorithm anyway? Why does it suddenly appear as though it is everywhere in our lives? In a simplistic form, one can understand algorithms as various mathematical models that are performed in a controlled fashion on data to present output in other desirable forms. The deployment of algorithms in the modern world has everything to do with the design and refinement of computers and their software. In the early days of computing, algorithms were implemented using machine code (a string of binary digits such as 1s and 0s) and were executed on the computer's hardware directly.

As the application of computers became better adjusted to real-life problems, today's algorithms serve a broad range of fields and diverse functions, which include "search (e.g., search engines), aggregation (e.g., news aggregators), observation and surveillance (e.g., government surveillance), forecasting (e.g., predictive policing), recommendation (e.g., music platforms), scoring (e.g., credit scoring), content production (e.g., robot journalism),

$$P_{like} \times V_{like} + P_{comment} \times V_{comment} + E_{playtime} \times V_{playtime} + P_{play} \times V_{play}$$

Figure 4.1 An early version of TikTok's content recommendation algorithm based on a report from *New York Times*

and allocation (e.g., computational advertising)" (Saurwein & Spencer-Smith, 2021, p. 223).

Notably, the term algorithm is not to be conflated with Artificial Intelligence (AI), which are computer systems designed to perform tasks that would normally require human intelligence (Christian & Griffiths, 2016). Nevertheless, some of the powerful AI systems that made their ways to the public discourse in recent days, such as ChatGPT, Bard, and Midjourney, all had algorithms involved in their decision-making and data-processing mechanisms.

Social media algorithms

As part of the Internet ecosystem, social media rely heavily on algorithms to perform their key features and functions. Depending on the platforms under scrutiny, specific social media platforms tend to deploy algorithms at varying levels of scope and depth. In general, there are three common areas in which algorithms are put into actions in the daily operations of social media: content curation and recommendation, content moderation, and advertisement allocation. The remainder of the section will briefly introduce the technical basis of these algorithms as well as how their functions are manifested in mainstream social media platforms.

To start, content curation and recommendation refer to the ability of algorithms to find and present relevant content to users on a given platform. This is necessary because the quantity of content being produced on social media is too mind-numbing for users to sort through in chronological order. An example of this is TikTok's "for you" page, which, according to the company ("How TikTok recommends," 2020), is generated for each user based on user interactions factors (like, share, comment, and follow), video information (captions, sounds, and hashtags), as well as users' device and account settings such as language, country, device type.

In the context of content recommendations, algorithms remain shrouded in secrecy, leaving users to form their own theories about what drives visibility on the platform. But are these theories accurate? In one study (Klug et al., 2021), researchers interviewed 28 TikTok users and identified three main criteria assumed to influence the algorithm: video engagement, posting time, and adding and piling up hashtags. They then analyzed tens of thousands of videos from the TikTok trending section to test these assumptions. Ultimately, it was uncovered that that higher engagement through likes, comments, and shares does indeed increase the likelihood of a video trending, as does posting at certain times. However, the belief that using trending hashtags and piling them up would significantly impact the algorithm was found to be false. Regardless, as users are trying to detect regularities and patterns of social media algorithms, there lies a conundrum, which is that, while the platform's algorithm governs users' behaviors, it is also informed by users' behaviors and how they infer about algorithm work.

However, it is also important to note that, on social media, content refers to more than just what users read and watch. It also includes recommendations for whom to connect with, what keywords to search for, and which pages to follow. In this way, content curation and recommendation make social media content more personalized and relevant for users.

If algorithmic content curation and recommendation optimize users' product experiences, content moderation then guards users against potentially harmful messages. Indeed, with the blowout of user-generated content on social media, recent years have seen an increasing amount of problematic content being widely circulated in the cybersphere, causing an immeasurable amount of economic loss and mental suffering. This content ranges from the all-too-common copyright infringement and hate speech to the invidious sexual-abuse content and horrific live-streamed massacres. Amidst the rising government mandates and public outcry for more effective interventions, social media companies have deployed various tech-based solutions. Chief to their toolbox is algorithmic moderation, which are "systems that classify user-generated content based on either matching or prediction, leading to a decision and governance outcome (e.g., removal, geoblocking, account takedown)" (Gorwa et al., 2020, p. 3).

From a technological standpoint, all algorithmic moderation systems help flag problematic content in one of two ways: matching and classification. Matching involves transforming a piece of content into a hash—a string of unique identifiers that is essentially a "digest" of the message. The system then compares that hash with a known library of hashes (which were deemed problematic) to see if the new content contains identical elements. Such a technique reduces the computing power needed to perform the content-moderation task and can be highly effective in preventing copies of a given content from being re-uploaded and circulated online. On the other hand, classification involves having the systems automatically identify the statistical patterns from a given content and put the content into several categories (e.g., offensive vs. no-offensive). Such a system is therefore more versatile in its application but demands more advanced software input to ensure the accuracy of the prediction.

As smart as the content moderation systems might sound, they sometimes produce mishaps. For instance, in 2020, fans of Felix Kjellberg — YouTube's biggest content creator at the time—accused the company of "shadowbanning" PewDiePie (Kjellberg's channel) by intentionally blocking some of his recent uploads. Although the company later clarified that this outcome was entirely due to YouTube's content moderation algorithm, some users hold the belief that mainstream social media companies are actively targeting certain content creators. Of course, users are hardly to blame in this case. As West (2018) rightly pointed out,

> content moderation systems remove content at massive levels of scale but do not do much to educate users about where they went wrong . . . the design of content moderation systems . . . makes people feel confused, frustrated. . . . The overwhelming reliance on flagging mechanisms to identify content to be removed reinforces a sense among users that platforms are a place where they can be targeted for their speech, beliefs, or identity.
>
> (p. 4380)

The third area in which algorithms are deployed on social media is the allocation of advertisements. Advertising is the lifeblood of the social media industry, with companies making billions by turning our attention and time into revenue. But how do these companies know who to target their ads to? The answer lies in the vast amount of data they collect on their users. For example, Spotify, the Swedish audio-streaming platform with many social features, compiles a long list of data about its users (see: https://support.spotify.com/us/article/understanding-my-data). Included in that list are a set of user labels conveniently named "inferences," which are inferred users' "interests and preferences based on your usage of the Spotify service and using data obtained from our advertisers and other advertising partners." And what exactly are these so-called users' "interests and preferences"? Based on the author's own Spotify data query, the list of labels includes one's entertainment spending, restaurant spending, and event ticket spending, among others (see Figure 4.2). The average user may not be aware of the extent of personal information that these platforms hold about us, but the race for more accurate advertising systems prompts legitimate concerns about "surveillance capitalism"(Zuboff, 2018), where users activities are monitored and used for commercial gain. However, the recent rulings by the European Union on Meta's ad practices

```
Inferences - Notepad
File  Edit  Format  View  Help
{
  "tags": [
    {
      "yr_qtr": "2018-09-30",
      "overall_spend_very_low": 0,
      "overall_spend_low": 0,
      "overall_spend_medium": 0,
      "overall_spend_high": 0,
      "overall_spend_very_high": 0,
      "overall_trans_very_low": 0,
      "overall_trans_low": 0,
      "overall_trans_medium": 0,
      "overall_trans_high": 0,
      "overall_trans_very_high": 0,
      "event_tickets_spend_very_low": 0,
      "event_tickets_spend_low": 0,
      "event_tickets_spend_medium": 0,
      "event_tickets_spend_high": 0,
      "event_tickets_spend_very_high": 0,
      "event_tickets_trans_very_low": 0,
      "event_tickets_trans_low": 0,
      "event_tickets_trans_medium": 0,
      "event_tickets_trans_very_high": 0,
      "entertainment_physical_spend_very_low": 0,
      "entertainment_physical_spend_low": 0,
      "entertainment_physical_spend_medium": 0,
      "entertainment_physical_spend_high": 0,
      "entertainment_physical_spend_very_high": 0,
      "entertainment_physical_trans_very_low": 0,
      "entertainment_physical_trans_low": 0,
      "entertainment_physical_trans_medium": 0,
      "entertainment_physical_trans_high": 0,
      "entertainment_physical_trans_very_high": 0,
      "entertainment_general_spend_very_low": 0,
      "entertainment_general_spend_low": 0,
      "entertainment_general_spend_medium": 0,
      "entertainment_general_spend_high": 0,
      "entertainment_general_spend_very_high": 0,
      "restaurants_spend_very_low": 0,
      "restaurants_spend_low": 0,
      "restaurants_spend_medium": 0,
      "restaurants_spend_high": 0,
```

Figure 4.2 Spotify's list of inferences based on the author's own data query performed in spring 2022

signal a significant shift, challenging the unchecked use of such algorithms by social media companies for ad allocation (Schechner & Dolby, 2023). These regulations underscore the need for greater transparency and user consent in the way these companies collect and use personal data for advertising purposes, potentially heralding a new era in digital privacy and data protection.

Are your phones listening?

We all have experienced it: on a hot summer evening, you reached the fridge for some orange juice only to find none left. You lament about this inconvenience with your roommate or family, and the next minute, an ad for orange juice pops up on the feed of your favorite social media platform. Or, perhaps you just saw someone wearing a pair of shoes you liked, and the next moment you turn to Amazon or some other eCommerce websites, there it is—that pair of shoes listed on the recommended product page.

Spooky stories like these are countless nowadays (if you cannot get enough of them, try to find more on the website *New Organs*, which collects first-hand accounts of similar events); hence, it is no surprise that many people have reached the conclusion that their phones must be listening. After all, the only item we carry 24/7 nowadays is our phones. The truth, however, is less exciting than many of us speculated. While there were multiple fringe cases about some apps turning on the mic of our phones without our awareness, in practice, such cases are outliers. Listening to users' conversations without their consent is not only illegal in most countries but makes little economic sense—just the amount of cloud storage and computing power needed to analyze speech makes that practice not at all cost-efficient and very much unnecessary.

The strikingly accurate and timely ads that many of us have experienced are, in fact, products of behavioral advertising, or algorithmically inferred predictions based on a slew of information that the platform has collected from us—over a long period. In other words, there is no need for companies to jump a few legal hoops to wiretap users; all they had about our behavior data are already decent enough to predict what we need at which time.

We must acknowledge that content curation and recommendation, content moderation, and advertisement allocation are by no means the only areas where algorithms become useful for social media platforms. Depending on their specific uses, algorithms can also fulfill purposes such as generating subtitles and translations for audio-visual content, rendering photos and videos, and detecting errors when platforms experience malfunctions. In that regard, to truly grasp the extent of their influence on society, one must keep a keen eye on the latest advancements in fields such as computer science, electronic engineering, and user-experience design.

Algorithm awareness

We have now grasped that algorithms can mold our daily social media activities through mediating, gatekeeping, and structuring what we see and interact with. In the meantime, because these algorithms function behind the scenes, it follows then to inquire about algorithm awareness among the general public. This understanding is crucial on at least two accounts: first, algorithm awareness, on the individual level, directly affects users' decisions regarding whether or not to sign up for a new social media service, and if so, how much personal data to share with the platform (Shin et al., 2022). Second, on the societal level, the digital gap between being algorithmically aware and unaware is potentially indicative of how divided citizens in a given society would be economically, politically, and even culturally in the future. In fact, scholars have argued that, in industrialized Western countries, the notion of the digital divide is less about people's access to the Internet and devices (the first level digital divide) than about the digital know-how and the benefits it brings to the users (the second level digital divide) (Gran et al., 2021). We will examine these issues more closely as we dive into Part III of the book, where we situate the function of social media in specific contexts.

But until then, let us answer the very question we just raised: To what extent are people aware of algorithms and their applications in social media? At the point of this writing, we do not really know. Because answers to this question will inevitably require a survey of a nationally representative sample, which in and of itself is a costly endeavor. In addition, researchers would also have to decide how comprehensive the survey questions need to be: Should one just ask a single "catch-all" question like "To what extent are you aware of social media algorithms?" or pose several survey questions that tap into users' experiences with all kinds of algorithms we discussed earlier? These are challenging methodological decisions researchers have to wrestle with.

Fortunately, there are some data that we can consult to draw our inferences. For instance, a survey in Norway found that nearly 41% of the population reported having no awareness of algorithms (Gran et al., 2021). Similarly, in the European Union, 48% of the population was unaware of what an algorithm is (Grzymek & Puntschuh, 2019). It is likely that algorithm awareness in the United States is similar to or lower than numbers from these European countries. Given these findings, it is important to educate the public about social media algorithms and their impact on society. Thus far, the European Union has taken major steps in this direction with initiatives like the European Media Literacy Week, which is designed to raise awareness about algorithms and the importance of media literacy. Similar efforts led by the government ought to be considered by the US and other nations.

Platform economy

Algorithms by themselves are just technical tools, and hence, not value-laden. But the same cannot be said about the social media and tech companies that design and deploy algorithms to amass power and influence. So, in this section of the chapter, we will take a step back and examine the broader context of the so-called platform economy.

"Platform economy" refers to the entire economic ecosystem revolving around the various kinds of digital frameworks and technologies designed for social and marketplace interactions. Based on the scopes and functions, Kenney and Zysman (2016) outlined the following types of digital platforms commonly seen in the current global economy.

First, platforms for platforms. These are the software infrastructures that constitute the digital space in which the modern digital economy functions. Examples include operating systems like Apple's iOS and Google's Android or the cloud computing services such as Amazon Web Services (AWS) and Microsoft Azure.

Second, platforms that make digital tools available online and support the creation of other platforms and marketplaces. Notable examples include GitHub, a digital repository for open-source software programs and a go-to place for tech engineers to share questions and skills; and Zenefits, an online marketplace for human resource tools.

Third, platforms mediating work. These include productivity-related sites and services like LinkedIn, Zoom, Lark, and Slack, and crowdsourcing platforms such as Amazon Mechanical Turk.

Fourth, retail platforms. Notable examples include all-encompassing e-commerce sites such as Amazon, Alibaba, and eBay, which host millions of vendors.

Fifth, service-providing platforms. Depending on the specific services, they could be travel-service platforms like Airbnb and Uber, content streaming platforms like YouTube, Spotify, and Twitch, financial platforms such as Kickstarter and GoFundMe, or conventional social media platforms such as Instagram, TikTok, and Snapchat.

It is important to note that the taxonomy of these platforms is loosely based on their respective scopes and primary functions. As technical innovations evolve, new platforms may emerge that do not neatly fit into these existing categories. However, a common thread among all platforms, regardless of their nature, is the presence of algorithms operating at various layers, serving specific functions such as error detection, content optimization, and communication coordination, among others.

> **Exercise**
>
> We've just discussed about five basic types of digital platforms. As time goes by, new services must have been invited that can be added to the list. As a practice, try to find a new example for each of these five platform categories. After that, pick two platforms of a different nature from these examples and try to answer the following questions: Who owns or controls the platform? How is the economic value created? How is the created value shared between the platform owners and its users?

Issues with big digital platforms

The main reason why we ought to pay attention to these digital platforms is because of their dominant market power. Virtually all the platforms mentioned earlier operate as part of a duopoly or, in some cases, even a monopoly in its niche market. Such a market position often leads to destructive consequences in many industries, but more specific to the digital platform economy, three issues are most salient: self-preferencing conduct, commission fees, and the lack of transparency (Bostoen & Mândrescu, 2020).

Self-preferencing conduct refers to a platform favoring its own services and product over other independent suppliers. In the digital platform economy, self-preferencing almost always involves some technical maneuvers. For instance, Amazon has long been accused of gathering data about the best-selling products on its own eCommerce platform, then manufacturing and promoting its own Amazon-branded products over other sellers (Petropoulos, 2021). In other cases, a platform can simply disfavor competitors' services by imposing strict user terms and conditions. For example, Apple Store was reportedly rejecting some third-party payment apps due to their potential conflict with Apple Pay.

Digital companies also charge commission or consignment fees (often based on the percentage of each transaction) on the content or services sold on their platforms. For example, Apple is known for charging a 30% fee for every transaction that happened via its App store, including those in-app transactions. This makes companies such as Epic Games—the developer behind the popular mobile game *Fortnite*—unsettling, since gamers make a large sum of in-game purchases in Fortnite each year. In a surprising move in 2020, Epic Games launched a new feature that allows its players to bypass the Apple store (and other app/gaming platforms) and pay Epic directly at a much lower rate, which, in turn, provoked Apple to pull off Fortnite in the App Store within hours of the update's appearance. The dispute eventually prompted both companies to go on court to seek resolutions (Owen, 2022).

If big players such as Epic Games cannot easily swallow the 30% fee cut, independent creators at other platforms paying higher fees/cuts to the platforms may find it even more challenging. On platforms like YouTube, which pays its content creators (otherwise known as YouTubers) mainly through advertising dollars, the parenting company Google can take up to a 55% cut of the ads revenues. Figure 4.3 illustrates some of the main income sources for typical YouTubers.

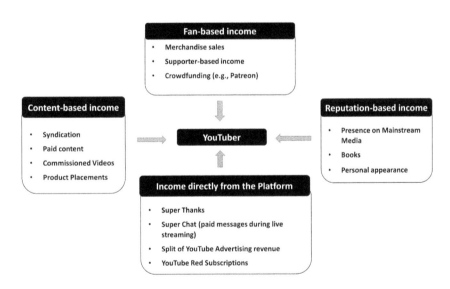

Figure 4.3 Sources of income for YouTubers

(*Source*: Figure redrawn partly based on Anable and Kenney (2017) as cited in Kenney and Zysman (2019, p. 31))

This led to the third issue facing most digital platforms: lack of transparency. Transparency is a multi-faceted term that concerns various stakeholders from a company's operational standpoint. It encompasses how platforms disclose their handling of user data, the platform's own decision-making process, the design and functioning of their systems, and how they deal with external requests for intervention and data access, among other things. Despite its complexity, users, creators, and developers have often experienced their content, accounts, or apps being removed or pulled off from various platforms without adequate explanations. Platforms often cite "violations of community guidelines" as a generic response to inquiries or appeals, leaving users stonewalled on how to avoid similar situations in the future. This lack of transparency puts platforms in an advantageous position in their interactions with users, potentially strengthening their dominant positions in the market.

Exercise

Increasingly, users and policymakers are voicing their concerns about social media companies' targeted advertising efforts, particularly regarding users' data privacy. To enhance its advertising and privacy transparency, Meta introduced two functions: a) a "Why am I seeing this?" button that provides users with an explanation of why they were shown a particular ad (ad explanations); and b) an Ad Preferences Page that provides users with a list of attributes Facebook has inferred about them and how (data explanations).

If you have not seen it, follow the instructions via this link (www.facebook.com/help/instagram/245100253430454) to see what Meta thinks about the topics you are interested in. Then, take a closer look at these labels. Do you find them to be accurate? Are there any labels that surprise you in any measure? What do you think about social media companies' efforts like this? Finally, is it helpful for you as an average user to understand what is happening "behind-the-scene," or is this yet another PR stunt to fulfill public's scrutiny?

Content producers

Different platforms form the digital infrastructure of the platform economy, while digital content represents the goods circulating within this infrastructure. Without content, there would be no values of any sort; hence, no users, no engagement, and no data. This makes content producers a vital group of users in the platform economy. By definition, content producers are the kinds of digital labor that produce much of what we consider digital culture products (e.g., videos, music, games, podcasts, and apps) and push them out through the platform. Platforms incentivize such productions by sharing with producers the profits earned from consumers' purchase of the digital culture product, virtual goods within the product (e.g., weapons or skins in games), and advertising income (Gawer & Cusumano, 2008).

As platforms are the ultimate determinant of the life of any given content produced by its creators, the rules and policies that govern their operations become a crucial aspect that digital content producers have to constantly navigate. An example of this is VidCon, an annual event initially founded by YouTuber John and Hank Green (ID: Vlogbrothers) as a gathering for YouTube content creators to showcase their work, exchange ideas, and seek collaborations. Over the years, as the so-called "creator economy" (or "influencer economy"; Chayka, 2021) catches up, VidCon has evolved into one of the world's largest content producers' conferences, attracting thousands of successful social media influencers and content creators to discuss their strategies for producing and circulating content on platforms such as YouTube, TikTok, Instagram, and many others. What is more, platform owners use the event to announce company updates and product launches, outline their rules and policies, and explain how they work with fans, content producers, and brands.

Through constant engagement with the platforms and peer learning, content creators aim to identity strategies and the best practices for content optimization in their own genre categories. For example, in an investigation of how digital content producers make sense of optimization strategies for their own cultural products, Morris and colleagues (2021) uncovered that, for musicians, optimization means tweaking their songs so that they are more likely to be featured on platforms' trending playlists. For mobile game developers, optimization tends to focus on tinkering with the game mechanics to ensure user engagement and retention or the monetization models that fit players purchasing habits. And app developers would have to choose to work on either longer-term subscription models or frequent and incremental updates. These efforts culminate in the competitive-creator economy that we live in nowadays. However, it is important to recognize that, under such an ecosystem, content creators often face exploitation, insecurity, and burnout. Moreover, despite their concerted efforts to circumvent algorithmic interventions, the enforcement of platform governance (such as content moderation and shadowbans) is experienced unevenly, depending on the creators' identities and content genres (Duffy & Meisner, 2022).

At times, the interplay among platforms, content creators, and average users may also introduce significant changes to traditional culture industries, so much so that the industry has to re-write its playbook and start anew. For example, on TikTok, one of the most popular short video sharing platforms, a plethora of creators publish content with musical elements in forms such as dances, lip-sync, music challenges, and audio/music memes, directly reaching global audiences in an instant (Toscher, 2021). Successful ones who can figure out TikTok's enigmatic algorithm stand a chance to quickly rise to public favor, bypassing the need to sign a record label. Further complicating the picture is that most creators who went viral on TikTok will soon release their music on Spotify, capturing even more music listeners. This pressed the music industry to scout TikTok for emerging talent and race to sign the next viral hit (Vox, 2022). Similar changes are also happening rapidly in the publishing, gaming, and even live entertainment industries, bringing new opportunities to professionals in the field while raising no small number of legal, financial, regulatory, and social questions.

The attention crisis

Our discussion thus far has been primarily centered on platforms/companies and content producers. In this final section of the chapter, let us turn to ourselves, the average platform users. As alluded to earlier, powered by the ever-evolving algorithms, today's social network sites have become increasingly capable of giving what users want—placing ads and recommending content that best fits our tastes and interests. From a business standpoint, it became conceivable why most social media would morph into what legal scholar Tim Wu referred to as "attention merchants" (Wu, 2017) whose chief purpose is to grab users' attention, which then gets monetized via advertisement sales. To be fair, social media companies are not the inventors of such an attention-driven business model; in fact, some of the harbingers in the newspaper industry, such as the *New York World* (and to a greater extent, the entire media industry in general), have long benefited from selling ads from users' continuous use/subscription of their services or products.

But one direct consequence of living in an attention-driven economy is that we are experiencing increasing difficulties in paying attention to things that matter most to us. By some estimation, college students now only focus on any task for 65 seconds, while workers only focus on average for three minutes at a time (Hari, 2022). So, if the fiercely competitive attention economy is not an original invention of social media, what has created this sudden boom of attention issues? Should we be legitimately concerned about the trend? And what can we do to reclaim our attention?

To start, and somewhat counterintuitively, we ought to acknowledge that, currently, there is no direct evidence linking the cause of social media use to the so-called attention crisis; for we can't go back in time and measure how the previous generations allocated their attention. However, inferences can be made through combining solid trend-research and individual experiences. In a study from *Nature Communications*, researchers (Lorenz-Spreen et al., 2019) used longitudinal datasets (covering multiple decades) across social media, books, and other Internet discourses and modeled the pace of the dynamic allocation of collective attention through the ebbs and flows of popular topics and cultural items. It was revealed that individual topics received shorter attention and higher turnover rates over the years. For instance, in 2013, a Twitter hashtag stayed within the top 50 for 17.5 hours on average, a number that gradually decreased to 11.9 hours in 2016. The researchers attributed the shorter attention cycles mainly to the increasing rates of content production and consumption. Anecdotally, the rate of people reporting any type of sleep deficiencies is also rising dramatically; yet, counterproductively, more and more people are trying to find refuges through "doomscrolling," a phenomenon of people spending an excessive amount of screen time for primarily negative news content (Ytre-Arne & Moe, 2021).

At the individual level, the impact of having too much information across a wide array of platforms is manifested through three mechanisms. First, the convenience and constant availability of smartphones, notifications, and new messages on our social network sites imply that social media users are highly likely to engage in "multitasking" or "task-switching"—the kinds of behaviors that often lead to limited task-performances. Multitasking is typically understood as engagement in more than one task within a given period of time. Multitasking may take three forms: dual-tasking, rapid attention-switching, and continuous partial attention (Wood & Zivcakova, 2015). In a study about the influence of tasking-switching in study settings (Rosen et al., 2013), researchers instructed students to observe and note how other students (ranging from middle school to university students) typically study at home on a minute-by-minute basis; afterwards, a questionnaire was completed by each observant. It was revealed that students, on average, switched to other tasks every six minutes, mostly due to the need to check their text messages and social media. In another study about social media use and academic performances, Lau (2017) surveyed undergraduate students' social media multitasking behaviors during the study period. Results found that social media multitasking to be negatively correlated with academic performance. Interestingly enough, using social media for academic purposes was not a significant predictor of academic performance as measured by cumulative grade point average. You just cannot study with social media on the side.

One explanation for the consistent adversarial effect of social media multitasking within study or work is the notion of limited capacity theory (Lang, 2000). The theory assumes that humans have a limited number of cognitive resources manifested in the working memory in any particular instance. In work or study settings, our cognitive resources focus on relevant words and pictures, build internal relationships among selected elements, and connect new understanding with prior knowledge in long-term memory. The rapid attention switch during multitasking, however, can overload our cognitive systems and leads to deficits in performance.

The current attention crisis also has relevance to the supply side of the information ecosystem, as social media can influence the types of information that users pay attention to in subtle but significant ways. In one experiment, researchers (Kätsyri et al., 2016) found that, when participants watched a series of news clips on television while also viewing a tablet with positive or negative tweets, the presence of the tweets decreased attention to the news. However, participants paid more attention to negative tweets than positive ones. This suggests that users have a bias towards negative social media messages when multitasking. With the widespread use of algorithmic content recommendation systems, there stands a great chance that machines may discover humans' innate interest in negative content and provide more of it to keep users entertained, rather than what is best for them.

The implications of the attention crisis are significant, as humans require periods of mind-wandering, dreaming, or simply doing nothing to rest and restore their attention (Williams et al., 2018). Moreover, research has consistently demonstrated that deliberate mind-wandering can enhance human creativity (Agnoli et al., 2018). However, the attention merchants vying for our attention on our phones and laptops have encroached upon these important moments of downtime, whether it be during sleep or idle time such as a subway commute. The degree to which we can rest and restore ourselves during these moments may impact our ability to handle other tasks effectively.

In response to the growing concern about the attention crisis, a range of artists, scholars, scientists, and tech ethicists have surfaced to demand more ethical operations from tech giants and social media companies. These individuals, including Cal Newport, Jaron Lanier, Jenny Odell, and Tristan Harris, offer concrete actions that people can take to regain control over their attention. While it's true that, as a society, we won't become more focused or brilliant simply by following their advice or switching off our phones, their ideas are worth considering in our own engagement with social media.

> **Ideas for taking control**
>
> If you are interested in learning more about what to do in taking back the control of your attention, a good starting point might be *The Center for Humane Technology*—a non-profit organization that

focuses on the broad concept of digital well-being. The site offers many suggestions for beginners interested in getting proficient in attention management. The following are a few ideas from them:

- Turn Off Notifications

    Red is a trigger color that instantly draws our attention. Reclaim your time by turning off notifications.

- Remove Toxic Apps

    Remove apps that profit off of addiction, distraction, outrage, polarization, and misinformation.

- Download Helpful Tools

    While we can't solve tech with more tech, there are some tools out there that can help (such as Flux, Flid, Insight Timer, etc.).

- Eliminate Outrage from Your Diet

    We vote with our clicks. Don't support sites that pollute our cultural environment with vitriol via clickbait and outrage.

- Follow Voices You Disagree With

    Social media serves us content we already agree with to keep us online longer, eroding our ability to engage with people who don't share our views. To solve problems from poverty to racism to climate change, we have to come together and expose ourselves to different perspectives.

- Be Compassionate

    Social media profits off hate and anger because it generates more engagement. Let's fight back with compassion.

- Set Boundaries

    We use our phones and news feeds from the moment we wake up to falling asleep and even in the bathroom.

- Fully Disconnect 1 Day Per Week

    Disconnecting can be a powerful way to reconnect with yourself and your loved ones. It's not only good for you—collectively, we can reduce time spent on social media platforms by 15%, impacting bottom lines.

- Remember the Positive

  If you receive 99 positive comments on a post and 1 negative comment, which do you focus on? Our survival-biased brains tend to focus on the negative, even after we turn away from our tech.

- Support Local Journalism

  Don't force your local newspaper to play social media's clickbait game. Support your local newspaper directly by paying for a subscription so that we can remain the customer, not the product. Democracy doesn't work without healthy journalism.

Keywords: Algorithm; content curation; content moderation; hash; matching and classification; adverting allocation; algorithm awareness; platform economy; self-preferencing conduct; commission fees; transparency; content producers; attention crisis; limited capacity model; doomscrolling; multitasking.

References

Agnoli, S., Vanucci, M., Pelagatti, C., & Corazza, G. E. (2018). Exploring the link between mind wandering, mindfulness, and creativity: A multidimensional approach. *Creativity Research Journal*, 30(1), 41–53.

Anable, B., & Kenney, M. (2017). *Understanding YouTube and YouTuber business models* [Unpublished manuscript].

Bostoen, F., & Mândrescu, D. (2020). Assessing abuse of dominance in the platform economy: A case study of app stores. *European Competition Journal*, 16(2–3), 431–491. https://doi.org/10.1080/17441056.2020.1805698

Chayka, K. (2021). What the "creator economy" promises—and what it actually does. *New Yorker*. www.newyorker.com/culture/infinite-scroll/what-the-creator-economy-promises-and-what-it-actually-does

Christian, B., & Griffiths, T. (2016). *Algorithms to live by: The computer science of human decisions*. Macmillan. https://doi.org/10.22316/poc/01.1.10

Duffy, B., & Meisner, C. (2022). Platform governance at the margins: Social media creators' experiences with algorithmic (in)visibility. *Media, Culture & Society*. https://doi.org/10.1177/01634437221111923

Gawer, A., & Cusumano, M. (2008). How companies become platform leaders. *Sloan Management Review*, 49(2), 28–35. https://doi.org/10.1109/emr.2003.1201437

Gorwa, R., Binns, R., & Katzenbach, C. (2020). Algorithmic content moderation: Technical and political challenges in the automation of platform governance. *Big Data & Society*, 7(1), https://doi.org/2053951719897945.

Gran, A. B., Booth, P., & Bucher, T. (2021). To be or not to be algorithm aware: A question of a new digital divide? *Information, Communication & Society*, *24*(12), 1779–1796. https://doi.org/10.1080/1369118x.2020.1736124

Grzymek, V., & Puntschuh, M. (2019). What Europe knows and thinks about algorithms: Results of a representative survey. *Bertelsmann Stiftung*, 1–38. http://aei.pitt.edu/102582/

Hari, J. (2022). *Stolen focus: Why you can't pay attention*. Bloomsbury Publishing.

How TikTok recommends videos #ForYou. (2020). *Tiktok.com*. https://newsroom.tiktok.com/en-us/how-tiktok-recommends-videos-for-you

Kätsyri, J., Kinnunen, T., Kusumoto, K., Oittinen, P., & Ravaja, N. (2016). Negativity bias in media multitasking: The effects of negative social media messages on attention to television news broadcasts. *PloS One*, *11*(5), e0153712.

Kenney, M., & Zysman, J. (2016). The rise of the platform economy. *Issues in Science and Technology*, *32*(3), 61.

Kenney, M., & Zysman, J. (2019). Work and value creation in the platform economy. In Steve Vallas *Work and labor in the digital age* (Vol. 33, pp. 13–41). Emerald Publishing Limited.

Klug, D., Qin, Y., Evans, M., & Kaufman, G. (2021). *Trick and please. A mixed-method study on user assumptions about the TikTok algorithm* [Paper presentation]. 13th ACM Web Science Conference 2021 (pp. 84–92).

Lang, A. (2000). The limited capacity model of mediated message processing. *Journal of Communication*, *50*, 46–71.

Lau, W. W. (2017). Effects of social media usage and social media multitasking on the academic performance of university students. *Computers in Human Behavior*, *68*, 286–291.

Lorenz-Spreen, P., Mønsted, B. M., Hövel, P., & Lehmann, S. (2019). Accelerating dynamics of collective attention. *Nature Communications*, *10*(1), 1–9.

Morris, J. W., Prey, R., & Nieborg, D. B. (2021). Engineering culture: Logics of optimization in music, games, and apps. *Review of Communication*, *21*(2), 161–175.

Owen, M. (2022). *Epic games vs Apple trial, verdict, and aftermath—all you need to know*. https://appleinsider.com/articles/20/08/23/apple-versus-epic-games-fortnite-app-store-saga—the-story-so-far

Petropoulos, G. (2021). A European Union approach to regulating big tech. *Communications of the ACM*, *64*(8), 24–26. https://doi.org/10.1145/3469104

Rosen, L. D., Carrier, L. M., & Cheever, N. A. (2013). Facebook and texting made me do it: Media-induced task-switching while studying. *Computers in Human Behavior*, *29*(3), 948–958.

Saurwein, F., & Spencer-Smith, C. (2021). Automated trouble: The role of algorithmic selection in harms on social media platforms. *Media and Communication*, *9*(4), 222–233. https://doi.org/10.17645/mac.v9i4.4062

Schechner, S., & Dolby, N. (2023). Meta's Facebook faces fresh threat to sending personalized ads in EU. *WSJ.com*. https://www.wsj.com/articles/metas-facebook-needs-consent-to-personalize-ads-eu-court-rules-6c705f18

Shin, D., Kee, K. F., & Shin, E. Y. (2022). Algorithm awareness: Why user awareness is critical for personal privacy in the adoption of algorithmic platforms?

International Journal of Information Management, *65*, 102494. https://doi.org/10.1016/j.ijinfomgt.2022.102494

Toscher, B. (2021). Resource integration, value co-creation, and service-dominant logic in music marketing: The case of the TikTok platform. *International Journal of Music Business Research*, *10*(1), 33–50. https://doi.org/10.2478/ijmbr-2021-0002

Vox. (2022). *We tracked what happens after TikTok songs go viral* [Video]. YouTube. www.youtube.com/watch?v=S1m-KgEpoow

West, M. S. (2018). Censored, suspended, shadowbanned: User interpretations of content moderation on social media platforms. *New Media & Society*, *20*(11), 4366–4383.

Williams, K. J., Lee, K. E., Hartig, T., Sargent, L. D., Williams, N. S., & Johnson, K. A. (2018). Conceptualising creativity benefits of nature experience: Attention restoration and mind wandering as complementary processes. *Journal of Environmental Psychology*, *59*, 36–45.

Wood, E., & Zivcakova, L. (2015). Understanding multimedia multitasking in educational settings. In *The Wiley handbook of psychology, technology, and society* (pp. 404–419). Wiley Blackwell.

Wu, T. (2017). *The attention merchants: The epic scramble to get inside our heads*. Vintage.

Ytre-Arne, B., & Moe, H. (2021). Doomscrolling, monitoring and avoiding: News use in COVID-19 pandemic lockdown. *Journalism Studies*, *22*(13), 1739–1755.

Zuboff, S. (2018). *The age of surveillance capitalism: The fight for a human future at the new frontier of power*. Hachette.

Part II

5 Social media and the self

Rachel was a curious young woman who loved to travel and explore the world. When she was in her early 20s, she lived in several different places in the United States, including Colorado, Upstate New York, Vermont, and North Carolina. One day, while on a trip in Central America, Rachel found herself without the means to buy a smartphone. Determined to stay in touch with her friends and family and to share her adventures with them, she decided to sign up for Facebook.

As she used the social media platform, Rachel noticed that all of her friends back in the US seemed to have perfect lives. They were getting married, having children, and landing high-profile jobs. This made Rachel feel left behind. But when she returned to the US, she met her boyfriend and eventually got engaged and married. She shared all the highlights of her life on Facebook and received many positive comments.

However, as time went by, Rachel started to feel like she could only post the happy moments of her life online. Even when she was traveling with her husband and they were fighting the entire trip, she only shared the happy moments. When she got pregnant, she shared the joy of having a baby but didn't talk about the difficulties of pregnancy. Eventually, Rachel's marriage fell apart, and she posted on Facebook about moving back home with her parents and her son. She received a surprising number of private messages from friends asking if she and her husband had split up. In these private chats, she learned that many of her friends had also gone through divorces or had split up with their partners, something she had no idea about. She realized, as though for the first time, that her friends were not as glamorous as they appeared on Facebook and that their lives were not so different from her own.

The story of Rachel, as told in an episode of the podcast *Hidden Brain* (Vedantam, 2017), is a familiar one. Every day, millions of people worldwide log into their social media accounts, posting personal updates, sharing ideas or news, searching for entertaining videos, or exchanging thoughts

DOI: 10.4324/9781003351962-7

Figure 5.1 Self-expression on social media
(*Source*: GoodStudio/Shutterstock)

and feelings. However, beneath all the dazzling content that people create, consume, and interact with, there lies a common thread; that is, much of what we engage with on social media is driven by self-expression, whether from ourselves or others.

Self-disclosure and self-presentation

Self-expression refers to the acts of sharing personal information with others. Depending on one's motives for sharing, self-expression can take one of the two forms: self-disclosure and self-presentation. Broadly speaking, self-disclosure can be conceived as actions that communicate personal information about the *factual self* to others, regardless of its impact on one's own public image; whereas self-presentation encompasses the communication activities that aim to influence the impressions formed by an audience about oneself.

Both forms of self-expression share some overlaps yet possess some distinctive features. Self-disclosure emphasizes sharing a factual self (a façade of the self that is consistent with reality); hence, it is often considered an essential process in forming and maintaining relationships. In contrast, self-presentation stresses the strategic goal of creating an image in front of an audience; as such, the information one uses to build such an image can be either factual or non-factual. But that is not to say that self-disclosure cannot serve similar goals as self-presentation. For instance, a college senior who has landed a job at a big-tech firm may want to share this information on Snapchat with their friends for his/her own image-building purpose. In doing so, s/he is disclosing factual information about herself while also projecting a positive image. In that scenario, the act of self-disclosure is tantamount to self-presentation.

The prism of *self*

The concept of self is a complex and multi-faceted one and one that is central to both psychology and communication studies. Scholars in these fields have debated the nature of self-concept and have come to the conclusion that it is a dynamic and ever-changing concept that is shaped by our social context. This has led to the identification of three key dimensions of self-concept: the actual self, the ideal self, and the ought self.

The actual self refers to the part of ourselves that we currently possess and can show to others. It is often used to define who we are in general and tends to be consistent across different contexts. In contrast, the ideal self is the version of ourselves that we hope to become, embodying our aspirations, dreams, and ambitions. It is something that we strive to achieve but may not yet possess.

The ought self refers to the aspect of ourselves that we think others expect us to be but have not yet achieved. For example, someone who wishes to become a highly altruistic person but currently prioritizes their own well-being over others.

Finally, there is the true self, or the real self, which is often what people mean when they say "be yourself." It is the sum of the hidden aspects of ourselves, possessing qualities that we may keep concealed from others due to uncertainty of how they will be judged. These hidden aspects may not be necessarily negative and are often hidden from the public due to societal norms. For example, a woman may appear to be timid and soft in a society with strong hierarchal norms, but her true self may be just as ambitious and free as anybody.

How do we know which aspect of the self is on duty in specific contexts? Well, we don't know. However, in psychological research, one can get a rough estimation through what is called the "Me/Not-Me" response task. The procedure of the task is described as follows:

> "Participants respond as quickly as possible as to whether each of a series of adjectives is self-descriptive, by pressing either a button labeled "Me" or another labeled "Not Me." The speed with which these responses are made is an indication of the relative accessibility, or readiness to be used, of the various concepts."
>
> (Bargh et al., 2002, p. 36)

Relying on this technique and through laboratory experiments, researchers have found that people tend to reveal more of the actual self in face-to-face interactions, whereas during interactions online via a chatroom, they were better able to express their true-self qualities (Bargh et al., 2002). Results like this were seen as evidence of the Internet's ability to allow people to express their true selves more freely.

How do social media facilitate self-expression?

Both self-disclosure and self-presentation can be found in offline face-to-face settings and the online environment. But certain characteristics of social media may amplify both forms of self-expression, making self-disclosure and self-presentation more salient in virtual communication than in face-to-face communication. And these characteristics include: anonymity, reduced information-richness, asynchronicity, multiple audiences, and audience feedback. (Many of these were touched upon in Chapters 1 and 3.)

First, social-media communication and in-person communication are clearly demarcated by anonymity. The anonymity offered by many social media platforms such as Twitter, Reddit, or Discord may allow people to be more open in disclosing themselves. This can be particularly useful for people who have been historically marginalized for unjust reasons. Sexual-minority men, for example, were among the earliest to adopt online communication platforms for dating and discussing homosexuality in the 90s. Similarly, those who have experienced certain mistreatments that are difficult to share in public might choose to do so privately. As an example, researchers (Andalibi et al., 2016) found that it is a common practice for some Reddit users to use "throwaway accounts" (temporary accounts that a user can create without any link to their primary Reddit identity) to disclose personal encounters of sexual abuse.

The second characteristic of social media that can facilitate self-disclosure is the reduced information-richness. On social media platforms, communication tends to convey less information than face-to-face communication due to the lack of nonverbal cues such as gestures, facial expressions, and voice tones. This lack of information richness is deemed instrumental at times. For example, as we will touch on in the next chapter, people seeking romantic relationships online may choose to disclose certain private information (such as age and sexual orientation) without the fear of being judged or embarrassed, unlike in an in-person scenario.

The remaining characteristics of social media (i.e., asynchronicity, multiple audiences, and audience feedback) tend to exert stronger impact on people's self-presentation behaviors than the first two. Asynchronicity, as discussed in previous chapters, allows users to communicate with others in either real-time (synchronous) or delayed (asynchronous) manner. More specifically, the non-real-time, asynchronous nature of many Instant Messaging services, for instance, offers users ample room to contemplate what to say and how to say it (e.g., word choices, use of emojis, length of the expression, etc.). Moreover, many platforms present multiple groups of audiences to its users altogether as "followers," "subscribers," or "friends"; as a result, what used to be separated connections in offline settings—friends, family relatives, acquaintances, co-workers, and even companies/brands—are "flattened" into a single virtual context. This phenomenon, known as context collapse (Marwick & Boyd, 2011), potentially propels people to manage their impressions in accordance with the common denominators of their network. As a result, users would often choose to post about positive stories over negative ones.

Lastly, the feedback that social media users receive from their audience in the form of likes, shares, and comments can provide valuable insight into how well they are presenting themselves. This feedback can create a positive feedback loop, as users who receive a steady stream of views and reactions may feel encouraged to continue sharing similar content. However, even when users' posts do not receive the feedback they were hoping for, this can still be used as information for future self-presentations. For instance, if a user shares a personal story on Twitter and it is not well-received, they may choose to alter the way they tell stories or avoid that topic in the future. This process of self-evaluation and adaptation based on audience feedback is known as impression management (Goffman, 1959) and it is a crucial aspect of how people present themselves on social media.

Why do users express themselves differently on different social media platforms?

Our discussion so far offers a broad-stroke understanding about how self-expression can be more conspicuous on social media than in the conventional face-to-face setting. But, as our experiences can attest, social media is not this one solid thing. Within the realm of social media, platforms and services offer features that vary drastically in each of the five dimensions discussed earlier. Hence, it is not a surprise that, as users, our self-expressions online are attuned to specific platform features and affordances.

One of the primary reasons why people might share varying pieces of information on social media is that they have different purposes in using each platform. These goals may include a desire for self-expression, relational development, social validation and approval, gaining social resources and information, benefitting others by sharing information, and managing one's identity (Bazarova & Choi, 2014). For example, a celebrity sharing information about a nonfatal cognitive disorder like prosopagnosia on Twitter might be seeking some sort of social validation, while someone sharing their experience as a terminal disease survivor on a Facebook cancer support group may be aiming to provide helpful health information (as discussed in Chapter 8).

The second factor that underlies users' self-disclosure behaviors across various channels is people's privacy concerns. Every social media platform has a privacy structure/design that can influence the amount and type of personal information that users are willing to share. For example, on an anonymized site such as Reddit, users may be more willing to share personal information in order to benefit others or receive social approval, while on a site like LinkedIn, users may be more restricted in the personal information they share due to the platform's emphasis on professional backgrounds.

To complicate the matter, sometimes users may decide to exercise their agency and creativity within the platform's existing privacy structure so as to create favorable terms for themselves. This is best manifested in the phenomena known as "Finsta"—fake or secondary Instagram accounts created by some users as an outlet to share private, emotional, low-quality, or indecorous content with their close friends. In comparison with people's public-facing/real Instagram accounts (i.e., Rinsta), Finsta accounts tend to be less carefully edited and appear to be more "off-the-cut" (Huang & Vitak, 2022). Through this reconfiguration, Finsta users override the privacy restrictions of an existing and widely-used social platform to create opportunities for more meaningful and reciprocal forms of social interaction.

In addition to expression goals and privacy concerns, network size and diversity are also crucial factors in considering the kinds of personal information one wishes to disclose on social media. In general, people tend to disclose less personal information on social media platforms with large and diverse networks, as it can be challenging to navigate multiple identities in front of a dynamic social group (Rui & Stefanone, 2013). This is, in part, why people with extreme views often gather on tightly-knit social network sites (e.g., 4chan and Truth Social), where their self-disclosure is less likely to be challenged. But, by and large, most people are aware of the negative consequence of excessive self-disclosure: it can lead to their partners feeling overwhelmed and unwilling to continue being confidants. As such, refraining from disclosing too much personal information on platforms with large, heterogeneous networks reflects common social psychology.

Many of the factors underlying individuals' self-disclosing behaviors across social media channels are identical for self-presentations on different platforms. However, in comparison to self-disclosure, social media self-presentation tends to be narrowly focused on strategically presenting an ideal self in front of one's social network; therefore, the affordance of cue-manageability is a hard-to-ignore factor for individuals in deciding which platform to use and how to use them. Cue-manageability, from the users' perspective, refers to the ability to present or conceal certain audio-visual cues about themselves in their online communication (Valkenburg, 2017). On social media, this is often achieved through features from the most basic textual/photo editing and filters to the advanced, specific video effects and temporal display of content (e.g., self-destructing messages).

Although the differences between these features can be subtle or even parochial by design, they tend to have an outsized influence on users' self-presentation behaviors on specific platforms. Consider the impact of social media's content ephemerality and persistence: a survey among adult social media users showed that people in the US rank Facebook and LinkedIn highest on content persistence; consequently, they also report a strong need to present a consistent self on these platforms (DeVito et al., 2017). The same cannot be said for Instagram and Snapchat, where their ephemeral or temporary message features allow more flexibility in self-expressions. But even between Snapchat and Instagram, users approach them differently. One study found that people who primarily use Snapchat tend to feel as though they are their true selves; namely, they feel comfortable on Snapchat when displaying qualities that they would normally not show to others. In contrast, those who are mainly active on Instagram tend to report that they need to show the best version of themselves, or the ideal self (Choi & Sung, 2018). Again, this shows that how a platform is designed greatly affect their users' self-presentation approach. Perhaps that is why Instagram decided to launch the "stories" feature soon after Snapchat first invented it.

Finally, users present themselves differently across platforms depending on whom they imagine their audience to be in respective media environments. However, the notion of "imagined audience" has a deep theoretical root. In his highly regarded book, *the presentation of Self in everyday life* (Goffman, 1959), sociologist Erving Goffman argued that people often selectively engage in self-presentation. That is, they present some aspect of themselves in a certain way to one audience group but different parts of themselves (in another way) to another. What happens on social media is that we, as users, conjure an inferred self through the minds of the imagined audience on each platform. One could be a collaborative employee on Zoom, a cheerful son or daughter on a family chatgroup, a cosmopolitan young professional on Tinder, or a fervent reader on Goodreads. Beneath the self we present on each platform, our self-presentation strategies are heavily influenced by who our intended audiences are, how we expect to be seen, and what we infer the audience wants to see.

Social media and identity development

In the long run, social media self-expressions, be it self-disclosure or self-presentation, deeply affect people's identity development (Subrahmanyam & Šmahel, 2011). Similar to the notion of selfhood, people don't have just one but multitudes of identities. Common identity labels include gender, nationality, racial ethnicity, sexual orientation, and occupation. Moreover, identities are born out of prolonged engagement in various social and cultural contexts, where norms of social interaction and self-expression are established and exercised. Social media, in that sense, is yet another space where all that happens.

But the role social media plays in shaping and developing personal identities is most noteworthy among two groups of population: youth and individuals from traditionally stigmatized groups. For young people, social media platforms provide a space for exploring, expressing, and manifesting their identities, resulting in the emergence of various Internet subcultures. These subcultures range from those that are socially meaningful, such as fitness culture, to those that are potentially harmful, such as youth vaping, and many that fall in between, such as hookup culture, competitive gaming culture, and influencer culture. Similarly, for traditionally stigmatized groups, social media can be both empowering and perilous for identity development. For example, a study found that, for many gay men, Grindr serves as a safe space for expressing their sexual identities, but the hyper-sexualization of the app also leads to concerns of "slut shaming." (Jaspal, 2017). Racial minorities may face similar dilemmas: during the most recent pandemic, racism and xenophobia toward people of East Asian descent increased in the US, and many Asian minorities used social media platforms like Facebook to express grievances about hate crimes and share coping strategies (Abidin & Zeng, 2020). However, in the same context, research also shows that users who have higher levels of trust toward their preferred social media tend to report increased perceptions of threats toward specific groups, such as China and Chinese people (Croucher et al., 2020). This suggest that social media plays a complex role in shaping personal identities, with both positive and negative effects, depending on the context and the group in question.

Social media and memory

How people construct and conceive their current identities is also related to autobiographical memories. Because people's current views about themselves result from what they recall about their past and how they recollect earlier selves and episodes. As the American writer William Faulkner powerfully puts it, "memory believes before knowing remembers."

Unlike the personal diaries we kept haphazardly stored at the bottom of the drawer, social media serve as a constantly updated personal journal, storing all the messages, photos, status updates, and videos we ever created. Research has shown that sharing personal information on social media can help us remember events more vividly. For example, in one study, researchers instructed participants to keep a diary of their autobiographical events for six consecutive days. And, during this period, participants were asked to use or not use Snapchat on alternate days. Results found that people recalled significantly more diary entries and used substantially more words during the Snapchat days compared with the non-Snapchat days (Johnson & Morley, 2021).

Notwithstanding the mnemonic effect of social media, it is not possible for us to remember every single item we posted on social media, as that would be too demanding for our brains. Instead, we rely on the retrieval cues provided by social media platforms to reconstruct our memories and experiences. That is in part why Facebook and other platforms decide to "remind" us of a party photo of ours taken ten years ago or a parking ticket received on this date of last year. Upon receiving those reminders, we don't relive those moments of life but experience a mixture of nostalgia, joy, or embarrassment through our memory. To test how social media retrieval cues affect our re-constructed memory, researchers asked participants to write about their recent life events in response to a set of cue words such as "birthday," "dog," and "friend" (Hou et al., 2022). Specifically, participants were asked to recount the event by updating their status on WeChat, or as if they were writing up a diary. After a week or two, they were invited back to the lab and, given a surprise memory, test either with the original cue words or not. It was found that, without those cue words, participants in the social media group recalled the memories more inconsistently than in the personal diary condition. This suggests that specific retrieval cues on social media play a role in our autobiographical memory-retention, hence facilitating the reconstruction of our identity.

Selfie

Of all the activities and actions people partake to record life events or express themselves, selfie is perhaps the most commonly accessible one. The term "selfie" bears two distinctive yet connected meanings: on the one hand, it denotes the photographic objects in which selfie-takers are being recorded in relation to something, someplace, or some specific perceptual state (that is, some kind of mood or feelings). On the other hand, taking a selfie is also an action, which entails several processes such as capturing, editing, sharing, viewing, and commenting on image. Seeing through this lens, what we observed as the selfie culture is a culmination of many layers of identity work intersecting with objects, events, and places.

> **From a youth subculture to the most famous selfie in the world**
>
> As with many behaviors we consider mundane nowadays, it is worth remembering that selfie-taking was once a hype among only tech-savvy teenagers who are seen as narcissistic and attention-seeking. But headlines of celebrities and high-profile politicians such as Miley

> Cyrus and the Obamas' engagement with selfies turned the behavior into a spectacle. In a symbolic announcement, the publisher of the Oxford Dictionaries named "selfie" as the word of 2013.
>
> Another pop culture event that signifies the acceptance of the selfie as a social norm was the 2014 Academy Awards ceremony. During the live-broadcasted event, comedienne and Oscar's emcee Ellen DeGeneres invited a surprising star-studded group shot with a self-focusing camera phone, resulting in "the most famous 'selfie' in the world," which led the social networking site Twitter to crash shortly after the selfie was posted on the comedienne's personal account and attracted heavy user-traffic.
>
> But the Oscars group selfie was not a spontaneous action after all. Subsequent media reports suggested that the selfie may have been a commercial act sponsored by Samsung, and DeGeneres had rehearsed for the selfie with the company's latest smartphone.

For some, selfie is a highly concrete form of self-expression. The image of the self, situated in a space presented in the observed condition, speaks to the creator's identity. For example, through a digital ethnographic work on #museumselfie, Kozinets and colleagues (2017) investigated museum selfies as a cultural phenomenon and an art form. Their analysis provided many types of selfies taken in museums, such as individuals interacting with the art objects, blended into art, mirror selfies, silly/clever selfies, contemplative selfies, and iconic selfies. The process is therefore seen as selfie-takers acting out the museum visiting experiences that give their identity its uniqueness and their life its meaning.

Although the specific manifestations of one's own faces/bodies open room for selfie-takers to be accused of narcissism, selfies can also be empowering in the proper context. The #nomakeupselfie, a social media campaign largely accredited to Charity Cancer Research UK, for example, leveraged users' altruistic motivations and performative needs to raise awareness for breast cancer and its research (Deller & Tilton, 2015). Later health campaigns like the #SmearForSmear (Cervical Cancer awareness) and the renowned ALS Ice Bucket Challenge all share similar features.

Whether we like it or not, selfies also demonstrate access and power. For example, a selfie with an influential politician taken at his/her fundraising party speaks loudly about the selfie-taker's power and access to certain scenes and places. Whereas the selfie taken on the battlefield (with dogs, as many Ukraine soldiers did during the Russian invasion in early 2022), the forefront of a protest, or a ruined city after a horrific natural disaster are often seen as a form of needed participation, resistance, or recording.

Figure 5.2 Ellen DeGeneres' group selfie taken during the 2014 Oscars ceremony

Today, selfie-taking is accepted in most situations, but in rare circumstances, selfies are just not appropriate by common standards. For instance, funerals are often seen as an occasion of solemnity. As such, former president Barack Obama joining a selfie with Denmark and UK's former Prime Ministers during Nelson Mandela's memorial service was seen as a break of social norms, leading people questioning their funeral etiquette (Meese et al., 2015).

Exercise

Have you ever shared your thoughts or experiences about a funeral on social media? If so, what was the nature of your post? Take a moment to scroll through Instagram or Facebook and observe how others are sharing about funerals and consider the social norms and etiquette associated with posting about funerals. Consider the dos and don'ts for sharing about funerals on social media.

Before we delve deeper into the topic of digital mourning in the following section, it's worth reflecting on our own thoughts and feelings about this practice. Is it appropriate to share our grief and mourning online, or are there certain social norms we should be aware of? How do we respond to others' posts about grief and mourning on social media? Should we "like" them, reach out to them personally, or simply scroll past?

Mourning and grieving on social media

As more and more conventional offline activities are taken place in the online world, so do our most private actions—mourning and grieving. The motivations, norms, and rituals might differ depending on whom people are mourning and grieving for on social media. Through in-depth interviews with people who have practiced social media mourning for their deceased friends or relatives, Moore and colleagues (2019) found that mourners are purposive in choosing with whom and what content they share on social media during bereavement. Some mourners simply use social media to share information with other friends or family; others wish to discuss the death of the person with a broader mourning community or to commemorate the deceased. For those who are mourning for someone deemed not to be a close friend or family, such as a musician, an actor, a writer, or a scholar, their social media mourning practices can be seen as the construction and management of their own identity in regard to the deceased. In a study of fans' reactions to the death of Michael Jackson, researchers found that fans who constructed a strong personal identity around MJ experienced hardship in reconciling with the singer's death, whereas those who had their identity built around the fan group tend to experience the mourning process in interaction with the other fans and reported fewer negative feelings compared to their counterparts (Courbet & Fourquet-Courbet, 2014).

But, perhaps unsurprisingly, in both studies, researchers have found that, for many mourners, prolonged use of social media could be associated with a slower resolution of someone's death, potentially impairing the mourners in going on with their own lives.

Complicating the issue even further, sometimes, the identity of the mourners also interferes with the specific mourning practices. When Paul Walker passed away due to a car accident, actor Vin Diesel (both of them were considered the lead actors in the *Fast and Furious* franchise) constantly posted on his own Facebook page, publicly expressing grief about Walker's death. This gesture implicitly invited fans of both actors to go on Vin Diesel's page and share their own emotions, creating a scene of online "mass mourning" (Klastrup, 2018).

And lastly, social media is sometimes used by people who are grieving to crowdfund for funeral expenses. Often, that is due to the high cost of mortuary rituals (Nova, 2019). In these cases, the success of the crowdfunding campaigns may be seen as a public relations event, although the mourners may not have intended it that way.

Keywords: self-disclosure; self-presentation; three types of self; context collapse; Finsta; cue-manageability; multiple audiences; identity; autobiographical memory; selfie; digital mourning.

References

Abidin, C., & Zeng, J. (2020). Feeling Asian together: Coping with #COVIDRacism on subtle Asian traits. *Social Media + Society*, 6(3). https://doi.org/10.1177/2056305120948223

Andalibi, N., Haimson, O. L., De Choudhury, M., & Forte, A. (2016, May). *Understanding social media disclosures of sexual abuse through the lenses of support seeking and anonymity* [Paper presentation]. Proceedings of the 2016 CHI Conference on Human Factors in Computing Systems (pp. 3906–3918). https://doi.org/10.1145/2858036.2858096

Bargh, J. A., McKenna, K. Y., & Fitzsimons, G. M. (2002). Can you see the real me? Activation and expression of the "true self" on the internet. *Journal of Social Issues*, 58(1), 33–48. https://doi.org/10.1111/1540-4560.00247

Bazarova, N. N., & Choi, Y. H. (2014). Self-disclosure in social media: Extending the functional approach to disclosure motivations and characteristics on social network sites. *Journal of Communication*, 64(4), 635–657.

Choi, T. R., & Sung, Y. (2018). Instagram versus Snapchat: Self-expression and privacy concern on social media. *Telematics and Informatics*, 35(8), 2289–2298. https://doi.org/10.1016/j.tele.2018.09.009

Croucher, S. M., Nguyen, T., & Rahmani, D. (2020). Prejudice toward Asian Americans in the COVID-19 pandemic: The effects of social media use in the United States. *Frontiers in Communication*, 5, 39. https://doi.org/10.3389/fcomm.2020.00039

Courbet, D., & Fourquet-Courbet, M. P. (2014). When a celebrity dies . . . Social identity, uses of social media, and the mourning process among fans: The case of Michael Jackson. *Celebrity Studies*, 5(3), 275–290.

DeVito, M. A., Birnholtz, J., & Hancock, J. T. (2017). Platforms, people, and perception: Using affordances to understand self-presentation on social media. In *Proceedings of the 2017 ACM conference on computer supported cooperative work and social computing* (pp. 740–754).

Deller, R. A., & Tilton, S. (2015). Selfies as charitable meme: Charity and national identity in the# nomakeupselfie and# thumbsupforstephen campaigns. *International Journal of Communication*, 9, 18.

Goffman, E. (1959). *The presentation of self in everyday life*. Anchor. https://doi.org/10.1215/08992363-8090145

Hou, Y., Pan, X., Cao, X., & Wang, Q. (2022). Remembering online and offline: The effects of retrieval contexts, cues, and intervals on autobiographical memory. *Memory*, 30(4), 441–449.

Huang, X., & Vitak, J. (2022). "Finsta gets all my bad pictures": Instagram users' self-presentation across Finsta and Rinsta accounts. *Proceedings of the ACM on Human-Computer Interaction*, 6(CSCW1), 1–25.

Jaspal, R. (2017). Gay men's construction and management of identity on Grindr. *Sexuality & Culture*, 21(1), 187–204. https://doi.org/10.1007/s12119-016-9389-3

Johnson, A. J., & Morley, E. G. (2021). Sharing personal memories on ephemeral social media facilitates autobiographical memory. *Cyberpsychology, Behavior, and Social Networking*, 24(11), 745–749.

Klastrup, L. (2018). Death and communal mass-mourning: Vin Diesel and the remembrance of Paul Walker. *Social Media + Society*, *4*(1). https://doi.org/10.1177/2056305117751383

Kozinets, R., Gretzel, U., & Dinhopl, A. (2017). Self in art/self as art: Museum selfies as identity work. *Frontiers in Psychology*, *8*, 731. https://doi.org/10.3389/fpsyg.2017.00731

Marwick, A. E., & Boyd, D. (2011). I tweet honestly, I tweet passionately: Twitter users, context collapse, and the imagined audience. *New Media & Society*, *13*(1), 114–133.

Meese, J., Gibbs, M., Carter, M., Arnold, M., Nansen, B., & Kohn, T. (2015). Selfies at funerals: Mourning and presenting on social media platforms. *International Journal of Communication*, *9*, 14. https://doi.org/10.4324/9781315688749

Moore, J., Magee, S., Gamreklidze, E., & Kowalewski, J. (2019). Social media mourning: Using grounded theory to explore how people grieve on social networking sites. *OMEGA-Journal of Death and Dying*, *79*(3), 231–259. https://doi.org/10.1177/0030222817709691

Nova, A. (2019). As the cost of dying rises, more families try crowdfunding for funerals. *CNBC.com*. www.cnbc.com/2019/12/07/more-families-are-turning-to-crowdfunding-to-pay-for-funeral-costs.html

Rui, J. R., & Stefanone, M. A. (2013). Strategic image management online: Self-presentation, self-esteem and social network perspectives. *Information, Communication & Society*, *16*(8), 1286–1305. https://doi.org/10.1080/1369118X.2013.763834

Subrahmanyam, K., & Šmahel, D. (2011). Constructing identity online: Identity exploration and self-presentation. In *Digital youth* (pp. 59–80). Springer. https://doi.org/10.1007/978-1-4419-6278-2_4

Valkenburg, P. M. (2017). Understanding self-effects in social media. *Human Communication Research*, *43*(4), 477–490. https://doi.org/10.1111/hcre.12113

Vedantam, S. (2017). Why social media isn't always very social. *Hidden Brain Media*. www.npr.org/2017/05/02/526514168/why-social-media-isnt-always-very-social

6 Social media for relationship management

Relationships are the essence of human existence. They shape who we are, how we think, and how we interact with the world around us. From the earliest moments of our lives, we seek connection and belonging with others. As we grow older, we continue to form new relationships and nurture existing ones, finding joy, support, and meaning in the connections we create. Today, in the age of social media, our relationships are taking on new forms, as we navigate digital platforms to communicate, share, and connect with others. In this chapter, we will explore the complex and evolving landscape of social media and relationships, examining the ways in which these technologies are transforming the way we connect with others and shaping our understanding of what it means to be in relationship with others.

The nature of relationship

Before delving into the role of social media in human relationships, it is crucial to understand the various types of relationships that exist, including those with colleagues, classmates, family members, romantic partners, and even virtual connections like those with virtual idols (e.g., Hatsune Miku, the Japanese anthropomorphic character). Each relationship is unique in terms of culture, norms, identity, and expectations, but as a whole, they reflect the relational diversity in our lives.

Parallel to the typology of relationships, relationships can be viewed through the lens of intensity and strength, resulting in the distinction between strong ties and weak ties. Strong ties, such as close friends and family members, are often characterized by their voluntary and reciprocal nature, resulting in support and mutual understanding. This makes strong ties a foundation of bonding capital, where individuals with similar backgrounds and interests come together to foster trust, cohesiveness, and a sense of belonging. However, many strong ties begin as weak ties, such as acquaintances that later evolve into close friendships. Despite the fact

DOI: 10.4324/9781003351962-8

that most of us have more weak ties than strong ties, weak ties can play a valuable role as bridging capitals, connecting individuals across different social groups and communities, and providing access to various resources (Putnam, 2000). Research shows that people are more likely to receive job referrals from weak ties rather than strong ties, both in offline and online environments (Rajkumar et al., 2022).

On a different spectrum, relationships can also be evaluated based on their directionality. For instance, if you are a fan of a high-profile musician, the relationship you have is mostly unidirectional, where the power and agency is unequal between the two parties involved. This type of relationship is ubiquitous on social media platforms like Instagram, where a user follows someone else, or YouTube, where a user subscribes to a channel, or Twitch, where a user watches a stream. On the other hand, a mutual or bi-directional relationship acknowledges the existence and value of the connection, with actions symbolizing the mutual recognition, such as close friends inviting each other to social gatherings or romantic partners exchanging gifts. On social media platforms like Twitter, bi-directional relationships are governed by the etiquette of "following" and "following back," where following back symbolizes mutual recognition.

Stages of online relationship

So, what roles do social media platforms play in relationships of various kinds? Naturally, the answer depends on what stage of the relationship is currently at. In general, most relationships would go through four typical stages: early idealization, development, maintenance, and sometimes, dissolution. The following sections will delve into the role of social media in relationships across each of the four developmental stages.

Early idealization

The initial stage of online relationships often involves some level of early idealization (Walther, 1996). Unlike offline interactions, online-mediated communication offers sparse communication cues, which leave room for imagination or idealization. In the context of online dating, potential partners interact with each other based on finite social cues. In such cases, surface characteristics such as one's hobbies, personal backgrounds (e.g., age, gender, racial ethnicity, etc.), and geographical locations are crucial for forming the relationship. Users who receive this information tend to imagine what the other person would be like, given their limited social cues. Moreover, the idealization process can be further enhanced through the fact that users on social media have some control over the types of messages they wish to present and when to present them. As a result, users who

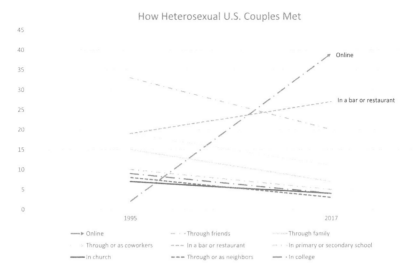

Figure 6.1 Changes in how heterosexual couples in the US met
(*Data Source*: Rosenfeld et al., 2019)

initially bond via social media tend to focus heavily on message production and curation, whether it is through editing profile photos or revising personal bios.

One space where early idealization happens most conspicuously is within matchmaking apps. For example, through a series of in-depth interviews with Tinder users in the Netherlands, Ward (2017) found that, on Tinder, where limited social cues were provided, users would engage in impression management primarily through photos and profiles. Experienced users would constantly contemplate and experiment on what to post and how to present themselves. In this process, they are quick to learn how to "decode" profile information to infer other users' socio-economic status and other information deemed essential for dating purposes.

Relationship development

As relationships progress from the initial idealization phase, they enter the development stage, where attaining intimacy becomes a key goal. This stage involves the process of social penetration, whereby individuals gradually reveal more about themselves to their partners. Self-disclosure plays a crucial role in this stage, but it is important to note that it is not done blindly. In fact, people weigh multiple goals to determine what and when to disclose. Derlega and colleagues (2008) conducted a study on college students with close relationships, where they inquired about the reasons for disclosing or withholding personal information from their relational partners (such as parents, friends, dating partners, etc.). Their findings revealed that trust maintenance, increased closeness, access to support, and obligation were reasons for disclosure, while fear of losing respect, protecting the other, and privacy were among the top reasons for withholding information.

The advent of social media has added layers of complexity to the process of relationship building. One concern is the idea that close relationships often rely on a level of information exclusivity. Sharing private matters only with those who are closest to us and vice versa is a crucial indicator of close relationships. However, if a close friend of ours regularly shares intimate information publicly on social media platforms, it can erode the connection between the two of you. Previous studies have shown that the more a friend self-discloses on Facebook, the lower the satisfaction and favorability associated with that relationship (McEwan, 2013).

To foster the development of newly formed relationships, private and personal correspondence through one-on-one channels is often more effective than public platforms. A study by Carpenter and colleagues (2018) randomly assigned college students to either send private messages to their group members or post information as a status update. The results showed that students who sent private messages reported significantly higher closeness ratings than those who posted status updates, indicating that private, targeted communication can play a crucial role in the development of relationships.

The evolution of relationships during their developmental stage often dictates the choice of communication platforms as they grow more intimate. Studies have shown that young adults, particularly women, tend to adopt a normative sequence of communication methods, starting with social networking sites for new acquaintances, then moving on to instant messaging, and finally transitioning to voice calls (Yang et al., 2014). The characteristics and perceived possibilities of different platforms can also affect their adoption and usage at different relationship stages. For instance, Snapchat, which is widely used among teens and young adults, is typically favored in more advanced relationships that aim to foster intimacy, rather than in the early stages of a relationship (Vaterlaus et al., 2016).

Relationship maintenance

Assuming a relationship successfully progresses through the development stage, it then enters the maintenance stage, where the individuals in the relationship work to maintain a certain level of trust and positivity. In contrast to the earlier stages of relationship development, the choices of communication channels during the maintenance stage become more diverse. The closer a relationship is, the more likely the individuals involved will adopt multiple communication channels. This phenomenon, known as media multiplexity/multimodality (Haythornthwaite, 2005), is specific to strong-tie relationships.

Maintaining a relationship requires attention to even the smallest details, as they can have a big impact on the overall health of the relationship. In fact, reciprocal interactions are a hallmark of mature relationships (Altman & Taylor, 1973). In the traditional offline world, this might involve inviting each other to events, sharing intimate information, and offering support. Online, this reciprocity takes many forms, from liking each other's Instagram stories to promptly responding to text messages. People often take into consideration both their relational strength and the features of the communication channels when deciding how to maintain their relationships.

However, achieving the right balance of reciprocity on social media can be challenging, as many factors can come into play. For instance, research has shown that women tend to expect their relational partners to acknowledge their social media posts through likes and favorites (Hayes et al., 2016), while in cross-sex relationships, males often comply with the norms set by their female partners (Yang et al., 2014). Additionally, the power dynamics between relational partners can also play a role, as those with less situational power tend to use more nonverbal forms of communication, such as text messages (Adams et al., 2018).

Emoji

And while we are on the subject of nonverbal communication, a trend in people's social media communication behaviors in the recent decade is the normalization of emoji in textual communication, particularly on social media. Originated in Japan, emoji is a form of digital pictogram designed to express emotions. Today, the notion of emoji has expanded significantly to include genres such as animals, places, objects, plants, weather, and various other symbols. Despite some underlying technical differentiations, many people use the term "emoji" interchangeably with emoticon, graphicon, or sticker (Tang & Hew, 2019).

The hidden aspects of emoji

Despite its Eastern root, the official entity that administers the library of emoji is a Silicon Valley nonprofit called the Unicode Consortium. A typical emoji character that wants to be enlisted in the library would undergo a formal review by the consortium's committee and, if approved, will receive a unified code, allowing it to be incorporated by all the tech companies. The particular aesthetic "look" of the emoji, however, can still depend on the proprietary font it represents; therefore, a typical smile face might look slightly different across Microsoft, Apple, Google, Facebook, and Twitter.

Today, with their wild popularity, emojis also became a form of branding and advertisement channel. And sometimes, companies would fight for the incorporation of certain emojis. For example, Taco Bell backed the creation of the taco emoji, Tinder supported the interracial couple emoji, and Ford Motor funded the application of the Blue Pickup Truck Emoji (Horowitz-ghazi, 2021).

Beyond going through the Unicode Consortium for new emoji approval, corporations are also attempting new ways of commercializing emojis. Apple's devices, for example, allow users to create "memoji" to match their styles and feelings, and the animated memojis also adopt users' voices and mirror facial expressions (Apple, 2022). Furthermore, in many Asian countries, stickers (emoji-like images, often offered as thematic sets) are rather popular in various messages apps and social network sites. Platforms such as WeChat and LINE often have their own specialized sticker stores, which are regularly updated and allow users to create and submit their own for public use, sometimes with or without financial compensation (de Seta, 2018).

Nearly anyone who has had experiences with emojis can agree that they are rather fun to use and can often help to facilitate communication between relational partners. Data from an online dating company Match.com suggested that users' frequency of emoji use is positively correlated with their frequency of dating and sexual engagement, indicating that emoji use is helpful in the context of romantic relationships (Gesselman et al., 2019). Beyond signifying senders' moods or feelings, people in close relationships use emojis to achieve various subtle communication purposes, such as managing a conversation (including initiating, maintaining, or ending a conversation), singling playful interactions (a sign of intimacy or closeness), and creating unique meanings (such as shared experiences or secrets) (Kelly & Watts, 2015).

Of course, emojis sometimes are often ambiguous in their meanings; therefore, depending on the situational settings, the use of certain emojis may need extra caution. Take the most popular emoji of 2021, 😂 (tear face with joy), for example. One might interpret it as a strong sense of

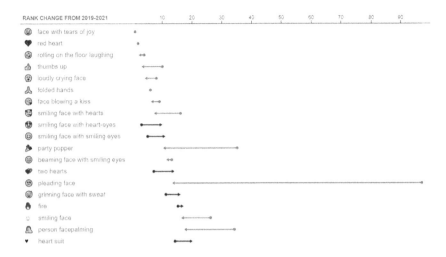

Figure 6.2 Emoji rank of 2021
(*Source*: Figure by Jennifer Daniel, Unicode Emoji Subcommittee Chair)

joy or as a reaction to embarrassment through laughter. However, things can be much trickier in a romantic relationship. For instance, a study found that in conflictual settings between romantic partners, using emojis dampens the partner's interest in the relationship, presumably because it conveys a lack of interest or seriousness in addressing the issue with the partner (Rodrigues et al., 2017). Yet, it is simultaneously true that using emojis may mitigate messages' negativity, hence serving as a way of de-escalation. For this reason, Stark and Crawford (2015) argued that "The patterns of use for emoji over time between friends and partners can become abstract and cryptic, or can degenerate to become *pro forma*: just plain basic" (p. 6).

Relationship dissolution

Whether we like it or not, relationships can end for a variety of reasons, such as changes in life circumstances, the emergence of new relationships, disapproval of behaviors, and physical separation. Social media can alter the process of ending a relationship by making it visible to both the relational partner and their imagined audience, which can shape how individuals approach the dissolution of the relationship. Depending on the visibility of the termination process on social media, it can be classified as a "soft" or "hard" process (Lopez & Ovaska, 2013). Soft processes involve actions such as unsubscribing, using specific friend lists to limit information access, or adjusting privacy settings, while hard processes include complete disconnection (unfriending/unfollowing), blocking, and/or reporting individuals to the platform. Regardless of the process chosen, these mediated relationship termination strategies afford users varying levels of ambiguity, which can help to alleviate the emotional and mental stress for those initiating the dissolution (Aoki & Woodruff, 2005).

Although we cannot precisely pin down social media as the cause of relationship failure, at times, they can be part of a problem that leads to a relationship derailing on two accounts. First, in virtually all healthy relationships, relational partners expect to communicate voluntarily and not to be coerced to communicate. Yet, as social network sites and instant messaging apps penetrate every dimension of our lives, they paved the way for a constant stream of Snapchat messages or Instagram photo tags. Under these circumstances, the relational partners on the message receiving ends might get bored, overwhelmed, or simply unable to keep up, leading to conflict or eventual withdrawal from the relationship (Bouffard et al., 2021).

The second aspect in which social media might contribute to relationship termination is that they make relational deceptions much easier. Here, we are not talking about the kinds of deceptions that are of criminal nature. Rather, we are focusing on the small and strategic lies people tell in specific relational settings. Certain features and affordances of social media might facilitate the production of deception such as the unlimited time constraints for composing texts, the capabilities to edit videos and photos, and the opportunities to upload unverified personal information (Toma et al., 2019). Ample research suggested that, in the online dating environment, deceptions can easily lead to relationship failure (Ellison et al., 2012). It should be noted that our discussion here does not imply that social media deception is much more prevalent than face-to-face communication. If anything, people are more likely to lie in record-less settings (e.g., in-person or via phone calls) than in other mediated channels such as social network sites and instant messages (Markowitz, 2022).

Needless to say, not all types of relationships can be dissolved in a clear-cut fashion. In romantic relationships, for instance, reluctant individuals might be surveilling or stalking ex-partners on social media: digging into their social media histories, deciphering every word or image, or constantly checking for personal updates. And as you might have expected, these social-media-surveilling activities are unhealthy, for they are often associated with more distressing and impaired recovery post-breakup. In fact, people who engage in social media surveillance tend to be the ones who have higher levels of relationship commitment and those who attribute the breakup to their partners (Fox &Tokunaga, 2015).

> **Exercise**
>
> Watch the Netflix crime documentary *The Tinder Swindler*. As you watch the show, pay attention to the media use and communication patterns between Simon Leviev, the antagonist of the story, and the three female victims. In particular, apply the concepts discussed in this chapter and analyze how social media use affects the relational dynamics during each stage of these relationships.
>
> After you complete this analysis, discuss with the class: To what extent do people's media use and communication patterns portrayed in this crime documentary differ from normal romantic relationships that you may have experienced?

Problematic use of social media in relationships

Most of what we discussed thus far regarding social media use in relational settings are situational; hence, the exact consequences of many behaviors can be hard to predict (think about the use of emojis). But there are certain behaviors, or more precisely, social-media-induced behaviors, that are almost categorically damaging to our connections with other people. One major type of those behaviors is phubbing. By definition, phubbing refers to the phenomenon of people staring at their phones instead of actively engaging in social interactions with others in offline settings. As smartphones become increasingly integral in people's daily lives, phubbing is seen everywhere and in almost all walks of life across the globe. In some settings, phubbing is often abrupt and unexcused and tends to cause interruptions to conversations or create a sense of interpersonal neglect.

For this reason, phubbees—victims of phubbing—might perceive phubbing as a form of ostracism. And indeed, one study found that people who recalled conversations being interrupted by unwanted phone phubbing reported a significantly higher sense of being ignored and being less valued (Hales et al., 2018), suggesting phubbing is of serious damage to relational partners. Interestingly, though, studies also found that phubbing can generate some sort of contagion: when one starts to do so, others will soon follow (Finkel & Kruger, 2012).

If phubbing can be harmful in relationships where relational partners are somewhat equal in their powers (e.g., friendships, romantic relationships), the consequences could be much worse in relational settings where the power dynamic is imbalanced. And that is precisely the case in the workplace. For example, boss phubbing seriously hinders employee engagement and dampens employees' trust in their supervisors. Employees also report being less valued by their supervisors, which is often associated with lower work performance and morale (Kanbur & Kanbur, 2021). In the family setting, phubbing by parents can result in poor parent-child relationships, insecure attachment, and negative emotions such as depression in adolescents (Zhang et al., 2021).

However, phubbing is not the only form of social ostracism that can occur on social media. Other actions, such as failing to like others' photos, delaying responses to instant messages, or unfriending someone, can also contribute to a sense of exclusion and harm relationships. For instance, one study found that not being tagged in photos on Instagram had a profound negative impact on people's satisfaction with their relationships and emotional well-being (Büttner & Rudert, 2022).

Social media in professional relationships

Professional relationships play a significant role in our social network, but we often don't give them much thought once we leave the workplace and enter our personal lives. In this final section of the chapter, we will explore how social media can play a role in building and maintaining professional relationships.

Social media can be an essential tool for establishing professional connections. Platforms such as LinkedIn offer job recommendations, discussions about job openings, opportunities to share professional experiences and advice, and the ability to make connections with individuals who would otherwise be unreachable for professional development. Indeed, one survey conducted among business-related graduate students found that the frequency of LinkedIn usage was significantly related to an array of networking benefits, such as career sponsorship, job search assistance, and protection and guidance (Davis et al., 2020). For companies looking to recruit employees, professional social networks can also be incredibly useful. In one study that investigated whether platforms like LinkedIn encourage dishonesty on users' resumes (Guillory & Hancock, 2012), participants were asked to create a resume on LinkedIn (where resumes are publicly available) and compare it with those who created a conventional offline resume. The results showed that people tend to be more honest about their verifiable job experiences on LinkedIn. Interestingly enough, people also tend to lie more often about their personal interests on LinkedIn—presumably due to the self-presentation needs in LinkedIn's publicly accessible online environment.

In the workplace, employees tend to have two types of relationships: instrumental and expressive ties. Instrumental ties are necessary for completing work-related tasks in formal work relationships (e.g., leader-subordinate and agent-customer relationships), while expressive ties are more informal, friendship-like connections in the workplace. Communication through enterprise social media platforms like Slack, Zoom, and Microsoft Teams is helpful for work engagement and can lead to higher productivity and job performance (Oksa et al., 2021).

With regards to expressive ties, social media can facilitate the formation of workplace friendships, which can contribute to employees' mental well-being and provide emotional and task-specific support (Wang et al., 2020). However, the positive effects of social media on expressive ties are stronger among those who have already established deep offline relationships. This may be why many American companies, after a prolonged period of remote work due to COVID-19, encouraged employees to return to the office, recognizing that in-person work relationships cannot be replaced by social media communication (Thomas, 2022).

> **Exercise**
>
> In the field of organizational and management studies, there is ongoing debate about the benefits and drawbacks of developing close friendships with coworkers. Some researchers argue that friendships within the workplace can foster cooperation and cohesion within teams or organizations. On the other hand, others claim that these friendships may lead to the formation of subgroups within the organization, potentially resulting in behaviors that conflict with the organization's goals (Pillemer & Rothbard, 2018).
>
> As college students who will soon enter the professional workforce, what are your expectations regarding your relationship with future colleagues? Do you wish to develop close relationships with your colleagues or do you wish for a clear boundary between you and your co-workers? Finally, what actions could you take on your own social media platforms to ensure you can achieve your relational goals in the work setting? Compile a list of ideas and compare yours with those of Ollier-Malaterre and colleagues' (2013, Table 1).

Keywords: types of relationships; strong and weak ties; relational directionality; stages of online relationships; relationship exclusivity; media modality; reciprocity; emoji; soft vs. hard relational-termination strategies; phubbing; social ostracism; instrumental and expressive ties.

References

Adams, A., Miles, J., Dunbar, N. E., & Giles, H. (2018). Communication accommodation in text messages: Exploring liking, power, and sex as predictors of textisms. *The Journal of Social Psychology, 158*(4), 474–490. https://doi.org/10.1080/00224545.2017.1421895

Altman, I., & Taylor, D. A. (1973). *Social penetration: The development of interpersonal relationships*. Holt, Rinehart, and Winston.

Aoki, P. M., & Woodruff, A. (2005, April). *Making space for stories: Ambiguity in the design of personal communication systems* [Paper presentation]. Proceedings of the SIGCHI Conference on Human Factors in Computing Systems (pp. 181–190). https://doi.org/10.1145/1054972.1054998

Apple. (2022). *Use Memoji on your iPhone or iPad Pro*. https://support.apple.com/en-us/HT208986

Bouffard, S., Giglio, D., & Zheng, Z. (2021). Social media and romantic relationship: Excessive social media use leads to relationship conflicts, negative outcomes, and addiction via mediated pathways. *Social Science Computer Review*. https://doi.org/10.1177/08944393211013566

Büttner, C. M., & Rudert, S. C. (2022). Why didn't you tag me?!: Social exclusion from Instagram posts hurts, especially those with a high need to belong. *Computers in Human Behavior, 127*, 107062. https://doi.org/10.1016/j.chb.2021.107062

Carpenter, J., Green, M., & Laflam, J. (2018). Just between us: Exclusive communications in online social networks. *The Journal of Social Psychology, 158*(4), 405–420. https://doi.org/10.1080/00224545.2018.1431603

Davis, J., Wolff, H. G., Forret, M. L., & Sullivan, S. E. (2020). Networking via LinkedIn: An examination of usage and career benefits. *Journal of Vocational Behavior, 118*, 103396.

Derlega, V. J., Winstead, B. A., Mathews, A., & Braitman, A. L. (2008). Why does someone reveal highly personal information? Attributions for and against self-disclosure in close relationships. *Communication Research Reports, 25*(2), 115–130. https://doi.org/10.1080/08824090802021756

de Seta, G. (2018). Biaoqing: The circulation of emoticons, emoji, stickers, and custom images on Chinese digital media platforms. *First Monday, 23*(9). https://doi.org/10.5210/fm.v23i9.9391

Ellison, N. B., Hancock, J. T., & Toma, C. L. (2012). Profile as promise: A framework for conceptualizing veracity in online dating self-presentations. *New Media & Society, 14*(1), 45–62.

Fox, J., & Tokunaga, R. S. (2015). Romantic partner monitoring after break-ups: Attachment, dependence, distress, and post-dissolution online surveillance via social networking sites. *Cyberpsychology, Behavior, and Social Networking, 18*(9), 491–498. https://doi.org/10.1089/cyber.2015.0123

Finkel, J., & Kruger, D. J. (2012). Is cell phone use socially contagious? *Human Ethology Bulletin, 27*(1–2), 15–17.

Gesselman, A. N., Ta, V. P., & Garcia, J. R. (2019). Worth a thousand interpersonal words: Emoji as affective signals for relationship-oriented digital communication. *PLoS One, 14*(8), e0221297. https://doi.org/10.1371/journal.pone.0221297

Hales, A. H., Dvir, M., Wesselmann, E. D., Kruger, D. J., & Finkenauer, C. (2018). Cell phone-induced ostracism threatens fundamental needs. *The Journal of Social Psychology*, *158*(4), 460–473. https://doi.org/10.1080/00224545.2018.1439877

Hayes, R. A., Carr, C. T., & Wohn, D. Y. (2016). One-click, many meanings: Interpreting paralinguistic digital affordances in social media. *Journal of Broadcasting & Electronic Media*, *60*(1), 171–187. https://doi.org/10.1080/08838151.2015.1127248

Haythornthwaite, C. (2005). Social networks and internet connectivity effects. *Information, Communication & Society*, *8*, 125–147. https://doi.org/10.1080/13691180500146185

Horowitz-ghazi, A. (2021, February 21). The story of the new blue pickup truck Emoji. In the indicator. *NPR*. www.npr.org/2021/02/15/968150292/the-story-of-the-new-blue-pickup-truck-emoji

Kanbur, E., & Kanbur, A. (2021). Phubbing at workplace. In U. Bingöl (Ed.), *#Trending topics on social media researches*. Peter Lang.

Kelly, R., & Watts, L. (2015). *Characterising the inventive appropriation of emoji as relationally meaningful in mediated close personal relationships* [Paper presentation]. Paper presented at Experiences of Technology Appropriation: Unanticipated Users, Usage, Circumstances, and Design, Oslo, Norway.

Lopez, M. G., & Ovaska, S. (2013, September). A look at unsociability on Facebook. In *27th international BCS human computer interaction conference (HCI 2013) 27* (pp. 1–10).

Markowitz, D. M. (2022). Revisiting the relationship between deception and design: A replication and extension of Hancock et al. (2004). *Human Communication Research*, *48*(1), 158–167. https://doi.org/10.1093/hcr/hqab019

McEwan, B. (2013). Sharing, caring, and surveilling: An actor–partner interdependence model examination of Facebook relational maintenance strategies. *Cyberpsychology, Behavior, and Social Networking*, *16*(12), 863–869. https://doi.org/10.1089/cyber.2012.0717

Oksa, R., Kaakinen, M., Savela, N., Ellonen, N., & Oksanen, A. (2021). Professional social media usage: Work engagement perspective. *New Media & Society*, *23*(8), 2303–2326. https://doi.org/10.1177/1461444820921938

Ollier-Malaterre, A., Rothbard, N. P., & Berg, J. M. (2013). When worlds collide in cyberspace: How boundary work in online social networks impacts professional relationships. *Academy of Management Review*, *38*(4), 645–669. https://doi.org/10.5465/amr.2011.0235

Pillemer, J., & Rothbard, N. P. (2018). Friends without benefits: Understanding the dark sides of workplace friendship. *Academy of Management Review*, *43*(4), 635–660. https://doi.org/10.5465/amr.2016.0309

Professional relationships: Guillory, J., & Hancock, J. T. (2012). The effect of LinkedIn on deception in resumes. *Cyberpsychology, Behavior, and Social Networking*, *15*, 135–140. https://doi.org/10.1089/cyber.2011.0389

Putnam, R. D. (2000). *Bowling alone: The collapse and revival of American community*. Simon and Schuster.

Rajkumar, K., Saint-Jacques, G., Bojinov, I., Brynjolfsson, E., & Aral, S. (2022). A causal test of the strength of weak ties. *Science*, *377*(6612), 1304–1310. https://doi.org/10.1126/science.abl4476

Rodrigues, D. L., Lopes, D., Prada, M., Thompson, D., & Garrido, M. (2017). A frown emoji can be worth a thousand words: Perceptions of emoji use in text messages exchanged between romantic partners. *Telematics and Informatics, 54*, 1532–1543. https://doi.org/10.1016/j.tele.2017.07.001

Rosenfeld, M. J., Thomas, R. J., & Hausen, S. (2019). Disintermediating your friends: How online dating in the United States displaces other ways of meeting. *Proceedings of the National Academy of Sciences, 116*(36), 17753–17758.

Stark, L., & Crawford, K. (2015). The conservatism of emoji: Work, affect, and communication. *Social Media + Society, 1*(2). https://doi.org/10.1177/2056305115604853

Tang, Y., & Hew, K. F. (2019). Emoticon, emoji, and sticker use in computer-mediated communication: A review of theories and research findings. *International Journal of Communication, 13*, 27. https://doi.org/10.1007/978-981-10-8896-4_16

Thomas, I. (2022). How companies are shifting their office spend to lure reluctant workers back. *CNBC.com*. www.cnbc.com/2022/06/04/how-companies-are-shifting-their-office-spend-to-lure-workers-back.html

Toma, C. L., Bonus, J. A., & Van Swol, L. M. (2019). Lying online: Examining the production, detection, and popular beliefs surrounding interpersonal deception in technologically-mediated environments. In *The Palgrave handbook of deceptive communication* (pp. 583–601). Palgrave Macmillan. https://doi.org/10.1007/978-3-319-96334-1_31

Vaterlaus, J. M., Barnett, K., Roche, C., & Young, J. A. (2016). "Snapchat is more personal": An exploratory study on Snapchat behaviors and young adult interpersonal relationships. *Computers in Human Behavior, 62*, 594–601. https://doi.org/10.1016/j.chb.2016.04.029

Walther, J. B. (1996). Computer-mediated communication: Impersonal, interpersonal, and hyperpersonal interaction. *Communication Research, 23*(1), 3–43.

Wang, B., Liu, Y., & Parker, S. K. (2020). How does the use of information communication technology affect individuals? A work design perspective. *Academy of Management Annals, 14*(2), 695–725. https://doi.org/10.5465/annals.2018.0127

Ward, J. (2017). What are you doing on Tinder? Impression management on a matchmaking mobile app. *Information, Communication & Society, 20*(11), 1644–1659. https://doi.org/10.1080/1369118x.2016.1252412

Yang, C. C., Brown, B. B., & Braun, M. T. (2014). From Facebook to cell calls: Layers of electronic intimacy in college students' interpersonal relationships. *New Media & Society, 16*(1), 5–23. https://doi.org/10.1177/1461444812472486

Zhang, Y., Ding, Q., & Wang, Z. (2021). Why parental phubbing is at risk for adolescent mobile phone addiction: A serial mediating model. *Children and Youth Services Review, 121*, 105873. https://doi.org/10.1016/j.childyouth.2020.105873

7 The use of social media among children and older adults

In the digital age, the influence of technology has reached even the youngest and oldest members of our society. While children may have formed unlikely bonds with virtual channels like Cocomelon, Chuchu TV, and Pinkfong, older adults are also adapting to life online, with many becoming avid users of social media and even going as far as becoming TikTok celebrities in their own right (e.g., Lillian Droniak @grandma_droniak, Chan Jae Lee @grandpachan). Despite their significance in our daily lives, children and older adults are often overlooked in discussions of social media and its impact on society. This chapter aims to shed light on the unique experiences and perspectives of these two important segments of the population.

Developmental stages of children

When it comes to the younger segment of the population, it is important to consider the development of children. Although minors, those under the age of 18, may not be fully responsible for their actions and decisions, most researchers differentiate between young children (under 13 years old) and adolescents (13 years and older). This differentiation is based on the significant cognitive development that occurs during childhood.

According to Swiss psychologist Jean Piaget (1936), young children up to 2 years old primarily focus on fundamental sensory perception, such as sights and sounds in their immediate surroundings. As they grow into the age of around 7 years old, they develop basic symbolic thinking, which involves rudimentary abstract concepts/symbols, language, and social relationships. Then, from age 7 to 11, more abstract concepts such as space, time, and mathematics begin to make much sense to them. And from 11 onward (up until 16), problem-solving skills such as logical and reasoning are strengthened, albeit far from fully developed.

From a slightly different perspective, developmental psychologist Lawrence Kohlberg (1974) outlined three phases for the development of moral reasoning in children. The first stage, pre-conventional moral reasoning, occurs in young children up to 9 years of age and is characterized by the interpretation of good and evil in terms of personal reward and punishment. The second stage, conventional moral reasoning, occurs from around 9 to 11 years of age and involves the acceptance of social rules regarding other people and social systems. The final stage, post-conventional moral reasoning, occurs as children gradually shift their focus to abstract principles and values that are applicable to all situations and societies.

It is fascinating to note that recent advancements in child psychology (Bloom, 2013) have revealed that moral development in humans begins as early as infancy. When capable of coordinating movement, babies will often attempt to soothe others in distress through patting and stroking. They also seem to have a rudimentary sense of justice, understanding that good actions should be rewarded and bad actions should be punished.

Given that many social media platforms have a minimum age limit of 13 and that we have covered teens and adolescents in Chapter 2, this chapter will focus on children who are 13 years old or younger, particularly those in or before the preschool ages.

YouTube for children

Children are using social media at a startling rate. According to a recent national poll administered by the C.S. Mott Children's Hospital (Clark et al., 2021), parents reported that half of the children between 10 and 12 years and one-third of children from 7 to 9 years use social media apps. And among all the social media apps, YouTube appears to be the most popular among young children: 80% of children between age 0 and 7 use YouTube, and 59% of them use YouTube Kids.

There are several reasons why YouTube is popular among children. The first is its ease of access and user-friendly interface. Prior to 2012, Google's YouTube app was pre-installed on virtually all android and IOS devices sold in the US market, granting many children easy access. (Apple discontinued pre-loading the app in 2012.) Moreover, the app's relatively simple home page layout allows young users to quickly access content through relevant content recommendations and search terms. Furthermore, YouTube videos often incorporate a range of different media, including audio, text, special effects, and animated characters. Such multimodal features can be particularly engaging and appealing for children, who are often visual and auditory learners.

The large variety of content available on YouTube is another factor that attracts children to the platform. Popular content and channels for children, such as nursery rhymes (e.g., ChuChu TV), animations (e.g., Peppa Pig, minions), unboxing videos (e.g., DisneyCollector channel), child vlogs (e.g., Ryan's World), and gaming (e.g., EthanGamer) can easily gain millions of subscribers. While these types of content may seem to cover different genres, researchers have found that they often share common features such as fast-paced scene changes, bright colors, and repetitive audio (Neumann & Herodotou, 2020a, 2020b). Again, these features may be particularly appealing to young children, considering their psychological development stages and the appeal of attention-driven movement and bright colors.

The popularity of YouTube content has prompted many to re-examine the potential negative impact of unregulated screen time on children. During the height of the COVID-19 pandemic, exhausted parents had little choice but to leave their kids with a tablet, sometimes for hours. To be fair, children were also asking for it, but as a result, screen addiction became a major concern. But the main issue is that unregulated screen time can also lead to a significant loss of free playtime and physical activities, which are considered detrimental to children's mental and physical development (Tandon et al., 2021). A recent study that followed nearly 900 children found that higher screen time (more than 1 hour per day) at the age of 2 was associated with lower scores in communication and daily living skills at the age of 4. On the bright side, researchers found that the frequency of outdoor play could help alleviate the impact of screen time on daily living skills, suggesting that outdoor play may mitigate the potential negative effects of screen time on developmental outcomes (Sugiyama et al., 2023).

Issues about child-oriented video content

The wide range of child-oriented content available online can also pose challenges to children and their families. Since content procedures on YouTube are spread across the globe and are not regulated under strict legal and ethical frameworks, many channels are not incentivized to create educational content like *Sesame Street*. Early childhood educators cautioned that excessive exposure to non-educational content might stall the regular timetable for children's development. Moreover, there has been an increase in disturbing content targeting children in recent years, such as the "Elsagate" controversy in 2017, which involved video content with inappropriate themes like violence, sexual situations, and toilet humor that was specifically targeted at children (Tahir et al., 2019). Despite efforts by YouTube to remove this type of content, it may still be accessible to some children due to content creators' efforts to bypass censorship through malicious labeling.

The potential risks of children being exposed to inappropriate or misleading content on online platforms are compounded by algorithms that may increase the likelihood of such exposure. One reason for this is that algorithms can be trained on biased data or content creators may manipulate the algorithm to promote their own viewpoints. For example, a child viewer who has expressed interest in content featuring extremely thin bodies may be recommended a range of content with unhealthy or unrealistic body standards, planting seeds for certain eating disorders in the future. Moreover, the deployment of algorithms in various areas may result in unexpected outcomes for parents. For instance, a study found that YouTube's algorithmic captions on videos for children mistakenly included inappropriate adult language in the text, such as "corn" being transcribed as "porn," "beach" as "bitch," and "brave" as "rape" (Ramesh et al., 2022). In their sample of over 7,000 videos from 24 top-ranked children's channels, almost 40% displayed words in their captions that were on a list of 1,300 "taboo" terms.

Another issue related to child-oriented video content on YouTube is the advertisement of products to children. Children under 13 are rapidly developing in their mental and physical abilities and are naturally drawn to food, play, and social interaction. Marketers are aware of this, and studies have shown that the most popular product categories targeted at children on YouTube are food, beverages, and toys (Coates & Boyland, 2021). Among advertisements for food and beverages, unhealthy options such as fries, cakes, and ice cream are more prevalent than healthier options. For example, one study in the US found that 94% of the content in a sample of 418 YouTube videos for kids contained unhealthy food and beverage advertisements (Alruwaily et al., 2020). Given children's susceptibility to advertisements, pediatricians and health organizations, including the World Health Organization, have warned that these ads may contribute to the global problem of child obesity. The issue is further complicated by the fact that many marketers collaborate with child influencers to promote their products, increasing the likelihood that children will consume unhealthy food. For example, a study found that influencer marketing of unhealthy food increased children's immediate food intake more than healthy options (Coates et al., 2019a). Unfortunately, traditional methods of regulation, such as labeling sponsored videos as advertising content, may not be effective for children. Research found that children who watched popular kid influencers' vlogs with advertising disclosure had a higher intake of the promoted unhealthy snack than those that promoted non-food products (Coates et al., 2019b).

Child influencers

The discussion thus far may leave readers with the impression that children are only passively accepting whatever social media are offering them,

The use of social media among children and older adults 103

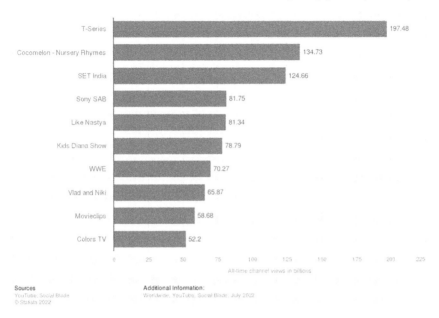

Figure 7.1 Three of the ten most viewed YouTube channels feature child influencers

but this is not the case. Aside from consuming social media content, children are also content creators. Like in the adult world, children who can produce content consistently and systematically can sometimes become real influencers, too, and even manage to amass millions of subscribers or followers. For example, three out of the top ten most viewed YouTube channels featured child influencers (Figure 7.1). And on overage, content featuring children and video games has three times as many views as other types of videos (Van Kessel et al., 2019).

However, the management of a large following across multiple platforms often requires adult assistance, particularly from parents. This becomes particularly important in the context of child influencers working with brands for product promotions, as adult input can provide crucial entrepreneurial insights (Craig & Cunningham, 2016). A prime example of this is Anastasia Radzinskaya, host of the YouTube channel *Like Nastya*, who, with the support of her parents, reportedly earned over $40 million in 2021, making her the highest-paid child influencer.

> **Exercise**
>
> In recent years, with the increasing prevalence of social media influencers as a normalized aspect of Internet culture, a subgenre of content known as "sharenting" has emerged (Plunkett, 2019). Sharenting refers to the practice of adult parents sharing private details about their child on social media platforms. Just like adult influencers, successful family influencers are sought after by brands for the purpose of product endorsement and direct marketing of merchandise to their followers.
>
> Have you personally come across any sharenting content on social media? (If not, go search your favorite platform for such content.) What's your view about them? Do you think parents should encourage their children to be influencers of any kind on social media? What might be some of the issues behind the phenomenon of sharenting?

The content produced by child influencers on platforms such as YouTube, Instagram, TikTok, and others tap into various forms and interests. At the point of this writing, popular genres of content among child audiences include unboxing videos, product/toy reviews, fun challenges, tutorials (mostly about games or makeup), personal vlogs, and performances. But what do all these mean for those child content creators, average child viewers/fans, as well as the attention-based social media economy as a whole?

First, we must admit that content production for children as content creators, on some level, is empowering. As children mature, socialize with others, and are exposed to the commercial world, they learn to behave and position themselves in ways they see fit. And for those who have strong ambitions online, social-media-content production is a powerful way to exercise their talent, construct their own identities, and sometimes, self-actualize. But empowering does not mean liberating. To establish and sustain that identity as "kidfluencers," children also have to put in tremendous amount of labor in producing content regularly, covering a wide range of topics and interests and interacting with fans in ways that can potentially increase the number of subscribers and followers. This is particularly palpable once we pay close attention to the communicative actions of the child influencers. For instance, many child influencers often describe their followers as "fans," ostensibly positioning themselves as celebrities (Martínez & Olsson, 2019). Some child influencers also adopt expressions, gestures, and behaviors that seem to emulate their adult counterparts (i.e., "smash the subscribe button!"). Such communication practices, in the long run, set up the norms and habits for children's online socialization (Abdul Ghani & Cambre, 2020).

For the vast majority of children as viewers, child influencers' content is often entertaining, interactive, and includes implicit messages for product promotions. This is partly because children are constantly seeking ways to learn, socialize, and bond virtually. And commercial products can be a way for them to relate to each other through consumer culture, market forces, as well as the broad social media economy. This means that child influencers may engage in activities such as playing with Lego toys, eating Kellogg's cereal, or visiting Disneyland in their content. By viewing this type of content, children can form online communities around products and influencers, create parasocial connections, and learn about related products.

As more and more children get involved in social media content production and consumption, the child-influencer economy becomes an established segment of the social media ecosystem in which viewers' attention is sold for advertising dollars and content producers are in constant race to stay relevant. Such a process of celebrification and commercialization also contributes to a peer-to-peer cultural industry (Marsh, 2016), where various cultural products are created, consumed, and transmitted. A few lucky ones, such as Justin Bieber and Alessia Cara, might become part of the celebrity culture, endorsing luxury brands, producing records, and attending award ceremonies. However, the majority of child influencers are ill-equipped to negotiate with powerful brands and platforms and are often exploited as a form of digital labor (Wright, 2019).

Law and regulations regarding child influencers

The growing trend of child influencers in the online world has drawn attention from legal scholars who are concerned about the power imbalance between these children and the commercial entities that benefit from their work. There have been calls for the creation of concrete legal protections for children in the context of social media and influencer marketing in order to prevent financial exploitation and ensure that children's rights are respected (Verdoodt et al., 2020; Masterson, 2020).

In the European Union, *the EU Directive on the protection of young people at work* provides a framework for the protection of children in the workplace. The Directive requires Member States to prohibit the employment of children under the age of 15 or those who are still in full-time compulsory education, with some exceptions for cultural, artistic, sports, and advertising activities, as well as certain types of light work by those aged at least 13 or 14. However, the Directive is limited in its scope, as it only applies to children who have an employment contract or an employment relationship under the law of the respective EU nation, and thus may not necessarily address the needs of all child influencers.

In the United States, the *Federal Fair Labor Standards Act* (FLSA) bans oppressive child labor, but its definition of "oppressive" is limited to forced physical labor, which is rare in today's context. Furthermore, the FLSA excludes media-related performance and entertainment activities from the definition of oppressive labor. As a result, the protection of child actors is largely based on state law, which varies widely across the country. This potentially can lead to dramatic differences in the level of protection for children depending on where they live.

In California—one of the most legislatively progressive states in the US—a recent attempt to extend the state's Child Actor's Bill to social media influencers was unsuccessful (Lambert, 2019). As a result, financial exploitation among child influencers remains unregulated. However, the state's recently implemented *Children's Online Safety Bill* does require tech companies to take proactive measures to protect young users (Singer, 2022), which could potentially protect child influencers from online harassment, including death threats, body shaming, and disability and identity-based insults, as well as the long-term psychological impact of such harassment (Gritters, 2019; Lorenz, 2018).

Virtual reality for children

The rapid development of the Virtual Reality (VR) industry has made VR gaming and headsets widely accessible to many children, making VR increasingly popular among children as well. Early industry surveys suggest that VR is quickly gaining popularity among children. For instance, in Fall 2016, approximately 40% of children (2–15 years old) in the US said they had never heard of VR. By Spring 2017, that number had dropped to 19%. A similar trend was also observed in the UK. With the growing ease of use and affordability of VR headsets, it is likely that a significant portion of children in the US and EU countries have had some exposure to VR.

But despite the sudden boom of the VR industry, there is limited research on the suitability of VR for children. Initial studies suggest that VR technologies may have benefits for children with certain types of cognitive disabilities such as Down's Syndrome or neurodevelopmental conditions such as Autism Spectrum Disorder (Freina & Ott, 2015; Mesa-Gresa et al., 2018). In addition, educational uses of VR among children are considered a promising area for hands-on learning, as VR experiences are often more interactive and vivid than traditional learning methods (Kamińska et al., 2019). However, there are also cautionary notes, as some studies found that VR may disrupt children's sensorimotor abilities, leading to an over-reliance on visual input over other sensory input (Miehlbradt et al., 2021).

Today, major VR companies, such as Apple, Meta and Microsoft, recommend their devices and products for users aged 13 and older, but it is common for VR devices to be purchased and gifted to children under that age limit. (We will discuss more about the impact of VR among the general population in Chapter 12.)

Exploitation in the virtual world

With the recent surge in popularity of the Virtual Reality industry, fueled in part by the buzz surrounding the metaverse concept promoted by Facebook's name change, a company specializing in gaming and VR has caught the attention of many people: Roblox. Roblox is a unique online platform where players can create and play a wide variety of digital games. Unlike traditional sandbox-style games like Minecraft, which typically offer a single, pre-defined virtual environment and storyline, Roblox offers a dynamic, user-generated experience that is particularly appealing to young children.

The platform is uniquely social and heavily centered around children. It is estimated that more than half of US kids and teens under 16 are on Roblox, and more than half of the platform's global users are under 13 years old (Clement, 2021). Investors of Roblox lauded the company for its effort in sponsoring workshops and summer camps to teach users skills of developing games within the company's platform, resulting in a wide range of user-created games that covers virtually all gaming genres, such as role-playing games, first-person shooting games, tower-defense games, and many others. From a business perspective, all games that are created by users are free to play, but creators could sell in-game digital goods to players, such as new skins, powers, weapons, or clothing.

In 2021, investigative video game journalist Quintin Smith published a series of two video journalism on the YouTube Channel *People Make Games*, criticizing Roblox for exploiting its child users. In these videos, Smith accused Roblox for taking lion shares of the revenues generated by game creators, intentionally setting a virtual-actual currency exchange system that plays in favor of the company, and charging a premium for creators who wish to push their games to players due to the platforms' poor discoverability. Beyond the alleged financial exploitation, many reports also indicate that the platform has issues with phishing scams aimed at intellectual property theft, ill-administered content moderation, and encouraging controversial micro-transactions among child users.

Older adults' use of social media

Aside from young children, older adults are another segment of the population that deserves special attention. According to a recent report from the United Nations (2020), there were approximately 727 million individuals aged 65 or over in the world, making up 9.3% of the global population. This number is expected to increase to around 16% by 2050. It is worth noting that the UN considers individuals over the age of 60 to be "older adults," although the term "elderly" is sometimes used to refer to individuals between the ages of 50 and 85 or older (Coto et al., 2017).

When it comes to social media use among older adults in the US, Pew research center (Auxier & Anderson, 2021) reports that 69% of adults between 50 to 64 and 40% of those aged 65 and older use some form of social media. In terms of specific platforms, it appears that Facebook and YouTube are the two most used services among people 65 and older (50% and 49%, respectively), while less than 10% of the population in this age cohort use Snapchat, Twitter, or TikTok.

There are several reasons why social media can be a beneficial choice for older adults. To start, a prominent advantage of social media is that it allows older adults to maintain close relationships with their loved ones and people with whom they have strong connection. As individuals age, they may prefer to spend more time with familiar people, and social media can facilitate this. Furthermore, social media can serve as a source of entertainment and companionship for seniors, who may have physical or health

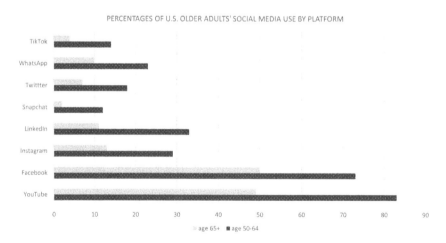

Figure 7.2 Percentages of US older adults' social media use by platform
(*Source*: Based on Pew Research Center's 2021 data)

limitations that restrict their ability to engage in certain activities. Platforms like YouTube and Facebook can provide a way for seniors to participate in enjoyable activities without leaving the comfort of their own home.

Second, older adults often have unique information needs related to healthcare, finance, and other subjects (Teng & Joo, 2017), and social media can be a valuable resource for meeting these needs. For instance, an older adult who is dealing with a chronic illness may join a Facebook group for individuals with the same condition, where they can share experiences and ask for advice from others who are going through similar challenges. They may also follow accounts on social media platforms that provide information and resources on healthcare, such as reputable organizations or healthcare professionals. Alternatively, older adults may use instant messaging applications to keep in touch with friends and family members who can provide guidance on specific tasks, such as financial planning or troubleshoot technical issues.

Third, older adults are at a high risk for experiencing depression and loneliness, and research has shown that social support from others can help mitigate the negative effects of loneliness on mental health (Park et al., 2013). In that regard, using social media to connect with others may provide seniors with the emotional support they need to reduce their risk of depression (Haris et al., 2014). What is more, there are evidences suggesting that using social networking sites like Facebook can improve senior people's executive function, which has connections to mental health and involves the ability to plan and execute tasks that require complex working memory (Myhre et al., 2017; Quinn, 2018).

Despite the benefits of social media use for older adults, many seniors remain on the sidelines of this digital revolution. While some of this can be attributed to a lack of basic access to the internet or appropriate technology, there are also those among the elderly who choose not to engage with social media. A study of older adults in Norway (Lüders & Brandtzæg, 2017), who intentionally abstain from using social networking sites, reveals that, while they recognize the potential value of these platforms, they often feel a sense of technical inadequacy, regarding themselves as too old to learn new technologies. Moreover, there is a perception among some older adults that social media is cold and narcissistic, which further fuels their aversion to these platforms. Privacy and information-security concerns also play a role, with some elderly individuals differing between sharing private information with commercial providers of social networking sites and losing control over who can access information within their personal network of friends and contacts. With Norway offering some of the highest living standards for the elderly, their experiences and perspectives offer valuable insight into how technology can be leveraged to improve the quality of life for seniors.

> **Exercise**
>
> Interview an elderly member in your family about his/her experiences using social media. Specifically, inquire about what platforms are they currently using. What kinds of content interest them the most? What opinions do they hold about social media and technology in general? Are there any encounters on social media that spark joy? Anything that agonizes them? If the person does not use any social media platforms at all, ask them why? And what would it take for them to sign up for a social media service?
>
> Once you complete the interview, compare your notes with a group of colleagues and offer a list of considerations/recommendations for social media designers so that they can create a better product/service for seniors.
>
> Finally, search your college library for an academic article by Hope and colleagues (2014) listed in the chapter's references. Then see how your recommendations differ or similar to the ones provided by these researchers.

Key Issues in older adults' use of social media

The use of social media by older adults also poses several complex yet challenging issues for society at large. Consider digital literacy, for instance. Digital literacy encompasses both technology and information literacy, and it plays a crucial role in determining an older adult's ability and comfort level with using social media. Technology literacy primarily concerns one's comfort and competency with devices and software use (or the lack thereof), be it using specific apps such as Teladoc for health visits or simply using smart TV for entertainment (Wang & Wu, 2021). In contrast, information literacy concerns the older adults' ability to find, evaluate, and acquire information for their needs. Lacking these skills can lead older adults to feel overwhelmed by the amount of information available on social media, leading some to withdraw from using these platforms altogether. Perhaps more damaging for those undeterred ones, a lack of information literacy can make older adults vulnerable to misinformation and scams, such as IRS imposter schemes and requests for assistance from fictive grandchildren. According to the Internet Crime Complaint Center (IC3) of the FBI, in 2022 alone, Americans lost more than $10 billion to these types of scams. And unfortunately, nearly half of the victims were over 60. For this reason, governments, NGOs, and local communities must consider improving digital literacy as a serious intervention for elderly population.

From a psychological perspective, older adults who are politically engaged tend to be less open to new thoughts and ideas compared to other populations.

This combination of characteristics can influence older adults' approach to social media in the sense that they may be more attuned to like-minded users and ideas, resulting in higher levels of political and network homogeneity (Kwak et al., 2018). Frequent contact with homogenous network on social media, in turn, also strengthen one's existing political beliefs, leading to strong in-group favoritism and, in some cases, political extremism. In fact, a study found that political polarization has been increasing rapidly among older adults, whereas a consistent depolarization was observed among young adults (Boxell et al., 2017). Chapter 9 will explore the consequence of political extremism and the extent to which social media plays a factor in political polarization.

To be fair, there are good reasons for some seniors to either avoid social media at once or stick together as a tight network online, as certain discourses on social media can be rather unfriendly and even hostile toward them. For example, one study found several conflictive generational discourses under the hashtag #BoomerRemover, referring to COVID-19 as "Boomer Doomer" or "Senior Deleter" (Meisner, 2021). Such discourses are disheartening, to say the least, and recovering our social media civility might be a first step to eliminating ageism.

Keywords: Child-oriented video content; kidfluencer; sharenting; celebrification; digital child labor; EU Directive on the protection of young people at work; Fair Labor Standards Act; technology literacy; information literacy; in-group favoritism; network homogeneity.

References

Abdul Ghani, M., & Cambre, C. (2020). "Ethan's golden YouTube play button: The evolution of a child influencer" chapter 4. In K. Warfield, C. Cambre, & C. Abidin (Eds.), *Mediated interfaces: The body on social media* (pp. 83–108). Bloomsbury Publishing. https://doi.org/10.5040/9781501356216.ch-004

Alruwaily, A., Mangold, C., Greene, T., Arshonsky, J., Cassidy, O., Pomeranz, J. L., & Bragg, M. (2020). Child social media influencers and unhealthy food product placement. *Pediatrics, 146*(5). https://doi.org/10.1542/peds.2019-4057

Auxier, B., & Anderson, M. (2021). Social media use in 2021. *Pew Research Center*. www.pewresearch.org/internet/2021/04/07/social-media-use-in-2021/

Bloom, P. (2013). *Just babies: The origins of good and evil*. Broadway Books.

Boxell, L., Gentzkow, M., & Shapiro, J. M. (2017). Greater internet use is not associated with faster growth in political polarization among US demographic groups. *Proceedings of the National Academy of Sciences, 114*(40), 10612–10617. https://doi.org/10.1073/pnas.1706588114

Clark, S. J., Schultz, S. L., Gebremariam, A., Singer, D. C., & Freed, G. L. (2021). Sharing too soon? Children and social media apps. *CS Mott Children's Hospital National Poll on Children's Health, University of Michigan, 39*(4). https://mottpoll.org/reports/sharing-too-soon-children-and-social-media-apps

Clement, J. (2021). *Distribution of Roblox games users worldwide as of September 2020*. www.statista.com/statistics/1190869/roblox-games-users-global-distribution-age

Coates, A., & Boyland, E. (2021). Kid influencers—a new arena of social media food marketing. *Nature Reviews Endocrinology, 17*(3), 133–134. https://doi.org/10.1038/s41574-020-00455-0

Coates, A. E., Hardman, C. A., Halford, J. C. G., Christiansen, P., & Boyland, E. J. (2019a). Social media influencer marketing and children's food intake: A randomized trial. *Pediatrics, 143*, e20182554. https://doi.org/10.1542/peds.2018-2554

Coates, A. E., Hardman, C. A., Halford, J. C. G., Christiansen, P., & Boyland, E. J. (2019b). The effect of influencer marketing of food and a "protective" advertising disclosure on children's food intake. *Pediatric Obesity, 14*, e12540. https://doi.org/10.1111/ijpo.12540

Coto, M., Lizano, F., Mora, S., & Fuentes, J. (2017). Social media and elderly people: Research trends. In G. Meiselwitz (Ed.), *SCSM 2017: Social computing and social media* (pp. 65–81). Springer. https://doi.org/10.1007/978-3-319-58562-8_6.

Craig, D., & Cunningham, S. (2016). Toy unboxing: Living in an unregulated material world. *Media International Australia, 163*(1), 77. https://doi.org/10.1177/1329878X17693700

Freina, L., & Ott, M. (2015, April). A literature review on immersive virtual reality in education: State of the art and perspectives. *The International Scientific Conference eLearning and Software for Education, 1*(133), 10–1007. https://doi.org/10.12753/2066-026x-15-020

Gritters, J. (2019). How Instagram takes a toll on influencers' brains. *The Guardian*. www.theguardian.com/us-news/2019/jan/08/instagram-influencers-psychology-social-media-anxiety

Haris, N., Majid, R. A., Abdullah, N., & Osman, R. (2014, September). *The role of social media in supporting elderly quality daily life* [Paper presentation]. 2014 3rd International Conference on User Science and Engineering (i-USEr) (pp. 253–257), IEEE. https://doi.org/10.1109/iuser.2014.7002712

Hope, A., Schwaba, T., & Piper, A. M. (2014, April). *Understanding digital and material social communications for older adults* [Paper presentation]. Proceedings of the SIGCHI Conference on Human Factors in Computing Systems (pp. 3903–3912). https://doi.org/10.1145/2556288.2557133

Kamińska, D., Sapiński, T., Wiak, S., Tikk, T., Haamer, R. E., Avots, E., Helmi, A., Ozcinar, C., & Anbarjafari, G. (2019). Virtual reality and its applications in education: Survey. *Information, 10*(10), 318. https://doi.org/10.3390/info10100318

Kohlberg, L. (1974). Education, moral development and faith. *Journal of Moral Education, 4*(1), 5–16.

Kwak, N., Lane, D. S., Weeks, B. E., Kim, D. H., Lee, S. S., & Bachleda, S. (2018). Perceptions of social media for politics: Testing the slacktivism hypothesis. *Human Communication Research, 44*(2), 197–221. https://doi.org/10.1093/hcr/hqx008

Lambert, H. (2019). *Why child social media stars need a Coogan Law to protect them from parents*. www.hollywoodreporter.com/business/digital/why-child-social-media-stars-need-a-coogan-law-protect-parents-1230968/

Lorenz, T. (2018, October 15). Instagram has a massive harassment problem. *The Atlantic*. www.theatlantic.com/technology/archive/2018/10/instagram-has-massiveharassment-problem/572890/

Lüders, M., & Brandtzæg, P. B. (2017). 'My children tell me it's so simple: A mixed-methods approach to understand older non-users' perceptions of social networking sites. *New Media & Society*, 19(2), 181–198. https://doi.org/10.1177/1461444814554064

Marsh, J. (2016). Unboxing videos: Co-construction of the child as cyberflâneur. *Discourse: Studies in the Cultural Politics of Education*, 37(3), 369–380.

Martínez, C., & Olsson, T. (2019). Making sense of YouTubers: How Swedish children construct and negotiate the YouTuber Misslisibell as a girl celebrity. *Journal of Children and Media*, 13(1), 36–52. https://doi.org/10.1080/17482798.2018.1517656

Masterson, M. A. (2020). When play becomes work: Child labor laws in the era of "influencers". *University of Pennsylvania Law Review*, 169, 577.

Meisner, B. A. (2021). Are you OK, Boomer? Intensification of ageism and intergenerational tensions on social media amid COVID-19. *Leisure Sciences*, 43(1–2), 56–61. https://doi.org/10.1080/01490400.2020.1773983

Mesa-Gresa, P., Gil-Gómez, H., Lozano-Quilis, J. A., & Gil-Gómez, J. A. (2018). Effectiveness of virtual reality for children and adolescents with autism spectrum disorder: An evidence-based systematic review. *Sensors*, 18(8), 2486. https://doi.org/10.3390/s18082486

Miehlbradt, J., Cuturi, L. F., Zanchi, S., Gori, M., & Micera, S. (2021). Immersive virtual reality interferes with default head–trunk coordination strategies in young children. *Scientific Reports*, 11(1), 1–13. https://doi.org/10.1038/s41598-021-96866-8

Myhre, J. W., Mehl, M. R., & Glisky, E. L. (2017). Cognitive benefits of online social networking for healthy older adults. *The Journals of Gerontology: Series B*, 72(5), 752–760. https://doi.org/10.1093/geronb/gbw025

Neumann, M. M., & Herodotou, C. (2020a). Evaluating YouTube videos for young children. *Education and Information Technologies*, 25(5), 4459–4475. https://doi.org/10.1007/s10639-020-10183-7

Neumann, M. M., & Herodotou, C. (2020b). Young children and YouTube: A global phenomenon. *Childhood Education*, 96(4), 72–77. https://doi.org/10.1080/00094056.2020.1796459

Park, J., Kitayama, S., Karasawa, M., Curhan, K., Markus, H. R., Kawakami, N., . . . Ryff, C. D. (2013). Clarifying the links between social support and health: Culture, stress, and neuroticism matter. *Journal of Health Psychology*, 18(2), 226–235.

Piaget, J. (1936). *Origins of intelligence in the child*. Routledge & Kegan Paul.

Plunkett, L. A. (2019). *Parenthood: Why we should think before we talk about our kids online*. MIT Press. https://doi.org/10.7551/mitpress/11756.001.0001

Quinn, K. (2018). Cognitive effects of social media use: A case of older adults. *Social Media & Society*, 4(3), 1–9. https://doi.org/10.1177/2056305118787203

Ramesh, K., KhudaBukhsh, A. R., & Kumar, S. (2022). 'Beach' to 'bitch': Inadvertent unsafe transcription of kids' content on YouTube. *arXiv preprint arXiv:2203.04837*. https://doi.org/10.1609/aaai.v36i11.21470

Singer, N. (2022). California governor signs sweeping children's online safety bill. *The New York Times.* www.nytimes.com/2022/09/15/business/newsom-california-children-online-safety.html

Sugiyama, M., Tsuchiya, K. J., Okubo, Y., Rahman, M. S., Uchiyama, S., Harada, T., Iwabuchi, T., Okumura, A., Nakayasu, C., Amma, Y., Suzuki, H., Takahashi, N., Kinsella-Kammerer, B., Nomura, Y., Itoh, H., & Nishimura, T. (2023). Outdoor play as a mitigating factor in the association between screen time for young children and neurodevelopmental outcomes. *JAMA Pediatrics.* https://doi.org/10.1001/jamapediatrics.2022.5356

Tahir, R., Ahmed, F., Saeed, H., Ali, S., Zaffar, F., & Wilson, C. (2019, August). *Bringing the kid back into youtube kids: Detecting inappropriate content on video streaming platforms* [Paper presentation]. 2019 IEEE/ACM International Conference on Advances in Social Networks Analysis and Mining (ASONAM) (pp. 464–469), IEEE. https://doi.org/10.1145/3341161.3342913

Tandon, P. S., Zhou, C., Johnson, A. M., Gonzalez, E. S., & Kroshus, E. (2021). Association of children's physical activity and screen time with mental health during the COVID-19 pandemic. *JAMA Network Open,* 4(10), https://doi.org/10.1001/jamanetworkopen.2021.27892

Teng, C. E., & Joo, T. M. (2017). Analyzing the usage of social media: A study on elderly in Malaysia. *International Journal of Humanities and Social Sciences,* 11(3), 737–743.

United Nation. (2020). *World population ageing 2020.* www.un.org/development/desa/pd/sites/www.un.org.development.desa.pd/files/undesa_pd-2020_world_population_ageing_highlights.pdf

Van Kessel, T., Toor, S., & Smith, A. (2019). A week in the life of popular YouTube channels. *Pew Research Centre.* www.pewresearch.org/internet/2019/07/25/a-week-in-the-life-of-popular-youtube-channels/

Verdoodt, V., van der Hof, S., & Leiser, M. (2020). Child labour and online protection in a world of influencers. In *The regulation of social media influencers* (pp. 98–124). Edward Elgar Publishing. https://doi.org/10.4337/9781788978286.00013

Wang, C. H., & Wu, C. L. (2021). Bridging the digital divide: The smart TV as a platform for digital literacy among the elderly. *Behaviour & Information Technology,* 1–14. https://doi.org/10.1080/0144929x.2021.1934732

Wright, M. (2019). YouTube 'kidfluencers' at risk of exploitation, children's commissioner warns. *Telegraph.com.* www.telegraph.co.uk/news/2019/09/20/youtube-kidfluencers-risk-exploitation-childrens-commissioner/

8 Social media for health and fitness

The COVID-19 pandemic has been one of the worst existential threats to humanity in the past 100 years. While the total death toll is devastating, the mental and economic toll of the extreme event is immeasurable and long-lasting. However, there have been some bright spots in the midst of this crisis: just a few months in, the virus's DNA structure was decoded, paving the way for the development of treatments and vaccines. Countries and communities around the world were able to quickly gather and distribute information about their infection rates, leading to effective interventions and responses. Finally, and perhaps empowering and disorienting at the same time, medical information and debate surrounding the virus were widely available to everyone through social media in near real-time.

Thanks to various digital communication technologies, there has never been a time in human history where health information has been so abundant and easy to access. But what are the benefits of using digital media for health information-seeking? Are there any risks or potential pitfalls associated with using social media for health and fitness? These are important questions for the well-being of any society, but before we delve into those, let's start by clarifying a few oft-used and closely-related terms.

According to the US National Library of Medicine, *health information* refers to a wide range of topics, including general health, drugs and supplements, specific populations, genetics, environmental health and toxicology, clinical trials, and biomedical literature. As such, *consumers of health information* include not only individuals with specific health conditions but also any individuals with health concerns and information needs, including the general public and caregivers.

Consumer health information, on the other hand, encompasses any information that enables individuals to understand their health and make health-related decisions for themselves and their families. This includes information that supports health promotion and enhancement, self-care, patient education, as well as the use of the healthcare system and the

DOI: 10.4324/9781003351962-10

selection of insurance or healthcare providers. Finally, *health information seeking*, as a specific form of health behavior, represents active efforts to obtain specific health information beyond the standard patterns of information exposure and use of interpersonal sources.

As alluded to earlier, with the ubiquity of social media and other Web 2.0 technologies, health information seeking is more diverse and convenient than ever. Statistics that follow offer a glimpse of just how widespread social media health information seeking is nowadays (Bryan et al., 2020; Tennant et al., 2015; Malani, 2018):

- Almost 90% of older users accessed popular social media sites to find and share health information.
- 54% of Millennials and 42% of all adults want to be friends with/follow their doctor on social media.
- 32% took a health-related action based on information they found on these platforms.
- 15% of parents with children under 18 have self-diagnosed a health concern based on information they found on social media.
- 68% of parents used social media for health information.

Appeals of using social media for health information

But convenience and availability aside, what else explain the value of social media for health? The answer is threefold: uncertainty reduction, emotional support, and peer support. For starters, when confronted with a health crisis, whether it's for ourselves or our loved ones, our initial instinct is to seek out information. This helps us assess the severity of the symptoms, determine whether to seek medical attention, and prepare for what may come next. Ultimately, this helps to alleviate the uncertainty surrounding the illness or condition. For example, in June 2022, Alexandr Wang, a young tech billionaire, tweeted the following:

> Had wisdom teeth surgery last Thursday, and experiencing a bunch of pain tonight on one side of my jaw. Still some reasonable amount of swelling. Should I be worried about this? Should I go get an X-ray ASAP to see if I've damaged a nerve? Is pain 6 days post surgery normal?
> (@alexandr_wang, 2022)

While this type of inquiry can be beneficial, one could argue that it often necessitates divulging personal health information. Indeed, research supports this link between the desire to reduce uncertainty and the willingness to disclose personal health information, with a survey of college students in the US, South Korea, and Hong Kong finding that those with a higher need for uncertainty reduction were more likely to disclose personal health information online across all three national samples (Lin et al., 2016).

In addition to seeking information, social media can also serve as a source of emotional support during times of health crisis. This can include sympathy, understanding, encouragement, and even physical affection (Moorhead et al., 2013). Emotional support can improve a person's mood and can be just as important, if not more so, as actual treatment for those suffering from mental, chronic, or terminal illnesses. But it is worth mentioning that emotional support can come from a variety of sources, including physicians, pastors, family members, friends, and even strangers. As such, the value of a particular source of emotional support can vary depending on the patient's age, health condition, stage of illness, and other factors. For example, an elderly person quarantined in a senior center during a pandemic might greatly value a video call from a family member, while an injured college athlete might find great comfort in reassurances of full recovery from their doctors through online consultations.

Finally, and taking its emotional support function one step further, social media can also provide peer support, particularly through online support groups, to patients and their loved ones seeking information and comfort from others in similar situations. This is important because we tend to place a higher value on interactions with those who are similar to us in some way, and this holds true in the healthcare sphere as well. To understand the types of messages being shared on social media, Myrick and colleagues (2016) analyzed a sample of tweets collected over a two-year period using the popular hashtag "#stupidcancer." They found that a large percentage (64.7%) of these tweets contained elements of information sharing and social support, and nearly 35% contained emotional expressions such as hope (the most common emotion), humor, anger, and fear (the least common emotion). Additionally, tweets about cancer with negative emotions were less likely to be shared than those with positive ones, suggesting that the cancer community on Twitter tries to maintain an optimistic outlook.

Issues about health-information seeking via social media

While social media can be a useful source of health information, it also presents challenges for those seeking credible information online. For one, health and medicine is a highly specialized field, and professionals typically undergo decades of training before advising patients. This means that patients and their family members with low health literacy may be vulnerable to poor or even inaccurate health information (Ventola, 2014). Inaccurate health information can also impact public health policies. For example, during the early days of the COVID-19 pandemic, information about injecting hydroquinone and hydrogen peroxide as a way to prevent the virus was widely shared on social media, leading to unnecessary suffering and medical waste during a health emergency.

Accessing medical and health information through social media may also lead to overreaction and overdiagnosis. For example, it has been reported that, starting in March 2020, hospitals and clinics in the US, Canada, Australia, and the UK saw an increase in teenage girls requesting a diagnosis for tics, a disorder characterized by physical jerking movements and verbal outbursts (Jargon, 2021). Doctors and clinical psychologists attributed this increase to the sudden popularity of TikTok videos featuring teenage influencers sharing experiences and diagnoses of Tourette's symptoms. And psychiatrists have labeled these symptoms, acquired from viewing illness-related content on social media, as "social media facilitated Factitious Disorder" or "social media associated abnormal illness behavior" (Giedinghagen, 2022). The treatment of such disorders is unlikely to be merely clinical but one that demands the collective effort of social media companies, schools, parents, and individual users.

One side effect of people's tendency to overreact to health-related information on social media is that it can be exploited by certain industries. The popularity of dramatic videos and photos showcasing the changes resulting from plastic surgeries on popular social network sites is a case in point. Research has shown that exposure to cosmetic-surgery-related information on social media can influence people's intentions and willingness to undergo cosmetic surgery procedures such as Botox injections or body augmentation (Zhao, 2022).

Exercise

Notwithstanding people's tendency to overreact to health information, some scholars argue that social media has a significant role in destigmatizing certain diseases and symptoms. For example, recent years have seen dramatic declines in discrimination toward people with mental health issues or autism spectrum disorder—in part, due to the normalization of discussing related subjects on social media. High-profile celebrities and influencers' disclosure of their own experiences also contributed to the wide acceptance of those issues among all walks of life.

Have you seen any discussions or content on social media that may have changed your views or perspectives on certain diseases or symptoms? Why do you think that this type of content is effective or persuasive in terms of destigmatization? On the flip side, have you seen any stigmatized conversations or content about certain diseases or illnesses on social media? If so, what would it take to change the stigmatized public discourse on those diseases?

Seeking health information through social media also raises privacy concerns regarding personal health information. In many parts of the world, personal health information is often considered extremely sensitive. Depending on its severity and social acceptance, discussing one's health problems can be considered undesirable in most offline settings. However, the anonymity of social media and its ability to bring together people with similar experiences offer users a safe space to disclose private health conditions and concerns. Unfortunately, this also means that any health information shared on social media is potentially subject to illegal abuse. Depending on how private health information is released to or accessed by third parties, some actions may violate the Health Insurance Portability and Accountability Act (HIPAA) Privacy Rule, while others may risk violating other federal laws.

The privacy risks associated with health-information seeking on social media are twofold (Libert, 2015). First, there is the risk of individuals' identities being publicly associated with their medical histories, which can lead to fraudulent activities such as false Medicare claims and identity theft. According to the latest data from the US Department of Health and Human Services (HHS), in 2021 alone, about 50 million Americans had their sensitive health data breached in 49 states (Leonard, 2022). The second privacy risk is the potential for unintentional discrimination. This can be particularly problematic in an increasingly algorithmic-based healthcare system. For instance, one study published in the journal *Science* revealed that a clinical algorithm widely used by hospitals to determine which patients needed care was exhibiting racial bias, such that Black patients had to be significantly sicker than White patients to be recommended for the same care (Obermeyer et al., 2019). This was because the algorithm had been trained on past data on healthcare spending, which reflects a history in which Black patients had less to spend on their healthcare due to longstanding wealth and income disparities. While this particular algorithm's bias was eventually detected and corrected, the incident raises the question of how personal information leaked online may be used in a way that accidently deepen the racial, gender, health, and/or socio-economic gap in the society.

Trends in social media healthcare

Our discussion on social media for health, thus far, has been centered primarily on individuals (as patients), but health as a formal social and business sector also encompasses other stakeholders, such as physicians, nurses, hospitals, government regulatory bodies, scientists, and even insurance companies. Later, we introduce a few industry trends that have immediate impacts on how patients receive healthcare through social media.

The first major trend that has concerned the entire healthcare industry in the past few years is the wide adoption of telehealth (a.k.a. telecare or telemedicine). Broadly speaking, telehealth can be defined as using information communication technologies, such as video teleconferencing, web-based programs, social media, and smartphone apps, for patient care. To be fair, delivering medical support through information technologies has been around since the 1960s. However, it is not until the recent decade, where the popularity of social media enabled physicians to network and share health information with their professional peers and directly interact with patients, that the telehealth industry actually took off. Take, for instance, the treatment of cancer patients, which is often multidisciplinary in nature and requires the patient to meet with numerous specialists on many occasions. Telehealth changed what was once a convoluted and heavily mediated communication between patients and doctors by removing the geological and temporal requirements, thus drastically improving the efficiency of the healthcare system. But, as telehealth becomes a normal part of healthcare, the industry also faces the challenge to figure out how to prevent physicians from overwork and burnout.

Perhaps somewhat ironically, one main situational factor that facilitated the rapid growth of telehealth across the globe was the COVID-19 pandemic, which forced patients and healthcare providers to seek ways to safely access or deliver health services (see Figure 8.1). According to the consulting company McKinsey, telehealth use had increased 38 times in 2021 compared with the pre-COVID-19 period (Bestsennyy et al., 2021). Among all medical specialties, psychiatry and substance use disorder treatment saw the most considerable growth in telehealth use. Of course, such a trend is inseparable from related legislation and regulatory changes. For

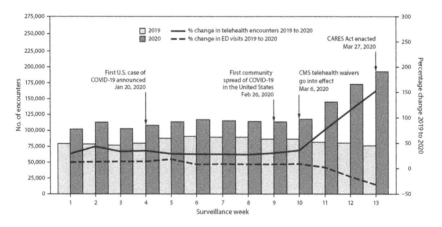

Figure 8.1 The number of telehealth patients surged during the early pandemic period

(*Source*: Koonin et al., 2020)

instance, during the pandemic, the US Centers for Medicare and Medicaid Services' expansion of reimbursable telehealth codes for the physician fee schedule incentivized physicians and insurance companies to work with patients virtually.

> **The dark side of telehealth**
>
> One company that rode on the recent tide of the telehealth boom was the subscription-based online mental health care company Cerebral. The company was founded in 2020, just before the worldwide outbreak of the coronavirus. As the need for online access to mental health services skyrocketed due to quarantines and lockdowns, Cerebral saw a surge of new users.
>
> One thing that made Cerebral particularly attractive to patients, aside from its virtual access, is that it can easily prescribe stimulants such as Adderall and Ritalin (both are heavily controlled substances under Drug Enforcement Administration in the US), without requiring patient examinations and visits. However, Cerebral was able to do so because, in the wake of the pandemic, the DEA temporarily suspended The Ryan Haight Online Pharmacy Consumer Protection Act of 2008, which is a law that makes it illegal to deliver, distribute, or dispense a controlled substance through the Internet.
>
> In addition to the regulatory change, the company also heavily engaged in social media advertising on platforms such as TikTok and Instagram and employed the esteemed gymnast Simone Biles as their chief impact officer. In early 2022, under heavy media scrutiny, the company halted prescriptions such as Adderall and Xanax for its patients. Government regulatory agencies such as the Federal Trade Commission also launched an investigation into the company's marketing conduct. Nonetheless, the case demonstrates that, without appropriate legal guardrails at side, telehealth can be corruptly exploited for financial gain rather than public interests.

Increasingly, hospitals and clinical facilities are recognizing the value of social media for patient communication. One study looked at nearly 300 hospitals' Facebook activities in the US and their connections with hospitals' ratings according to the Hospital Consumer Assessment of Healthcare Providers and Systems (Richter & Kazley, 2020). Results show that hospitals that were deemed active on Facebook not only received a higher satisfaction score but had more patients indicating a willingness to recommend them to others.

However, it is one thing to say that hospitals' social media presence contributes to patients' satisfaction; it is completely another for the hospitals to communicate with patients effectively on social media. To better understand what hospitals could do to improve their social media presence, De Las Heras-Pedrosa and colleagues (2020) conducted a focus group with representatives from various national and regional patient associations in Spain—a country with one of the most efficient healthcare systems in the world. Participants reported that they appreciate hospitals' effort in communicating with patients via social media but expected hospitals to post information that can be easily comprehensible and interactive (e.g., incorporating images and videos), rather than being overtly technical and jargon-filling. In addition, participants also expressed the need for hospitals to emphasize providing information that is relevant to healthcare rather than for marketing purposes. All of these suggest that health providers must adapt to the challenges and opportunities presented by social media and consider patients' communication essential to their daily operations.

Finally, thanks to social media, the use of big data for predicting and tracking diseases has now become possible and getting more accurate by the day. As discussed earlier, patients nowadays frequently turn to social media to search for and discuss health symptoms and conditions, and that resulted in a large amount of health data that is both lively updated and rich in its metadata (information about information), such as geolocation, time stamp, as well as other users' information. These data, in the form of search-engine keywords, social media posts, comments, photos, videos, and even likes and shares, allow data scientists and medical professionals to build mathematical models that can predict and monitor outbreaks of certain diseases, sometimes down to certain cities, districts, even street blocks. At present, researchers have developed and deployed such a model to surveil and predict the trajectory of outbreaks from normal influenza to the infectious virus, such as Zika virus, coronavirus, and monkeypox, with reasonable accuracy (Rousidis et al., 2020). As the sources of data increase and the models are continuously perfected, governments and societies across the world can expect social media-based disease prediction to become part of the formal public-health-monitoring system.

Social media for fitness and exercise

Aside from utilizing social media for specific health-related purposes, many healthy individuals turn to it for preventive healthcare, particularly fitness and exercise. A trend that embodies this is "fitspiration," where images and videos are shared on social media to inspire people to live a healthier lifestyle through exercise.

Two perspectives might facilitate our understanding of the appeal of fitspiration. From the content creators' perspective, fitspiration can be conceived as a form of social media self-presentation, where users project an idealized self-image by showcasing attractive lifestyles involving fitness and healthy food. On the other hand, social validation metrics such as likes, shares, and views act as incentives for users to keep creating similar content by actually engaging in exercise more frequently. From the viewers' perspective, consuming fitspiration content can be a source of motivation, which is best explained by the notion of up-wards social comparison, where users are compelled to act (exercise) because of wanting to achieve the lifestyle or body images presented by others (Robinson et al., 2017).

While fitspiration is intended to have a positive influence, its actual impact is often questionable. A content analysis of 600 images collected from Instagram utilizing the hashtag #fitspiration revealed that the majority of fitspiration images featuring female bodies portrayed thin and toned bodies (Tiggemann & Zaccardo, 2018). This raises concerns about the potential dissemination of false body norms, as these thin, idealized images may lead to dissatisfaction with one's own body, even among individuals who are otherwise considered healthy (Burke & Rains, 2019). An example of this is the once-trendy body-checking videos on TikTok and other social media platforms, where users showcase their hourglass figures in front of the camera. And, parallel to the female fitspiration, approximately 30% of those images featuring male bodies displayed a high degree of muscularity. As with the thin body image for women, research has also linked exposure to muscular ideals with body image decrements for men (Tiggemann et al., 2007).

Self-tracking

As the technology, social media, and consumer healthcare industry evolve, a unique self-tracking market has emerged (Neff & Nafus, 2016). Self-tracking is the act of using various technologies to record details about one's body and life, often for the purpose of gaining self-knowledge, self-reflection, and self-improvement. There is now a plethora of products and services available that offer users a "quantified self," including fitness-tracking and calorie-counting apps (e.g., Runner Keeper and Strava), as well as hardware such as smartwatches and smart fitness mirrors.

To be clear, humans have been performing health-related self-tracking for millennia but today's technology and social media offer a unique blend of continuous, real-time monitoring, wireless data transfer, and cloud-computing storage that reduces costs and the need for expertise. These technologies also allow users to combine different data sets to identify patterns in a way that was previously impossible, leading to potentially more personalized health and medical recommendations. Moreover, the visually appealing presentations generated by self-tracking software and services conform to the concept of the "data spectacle" (Gregg, 2015), where users of self-tracking technologies are motivated by the opportunity

124 *Part II*

Figure 8.2 The Quantified Self movement
(*Source*: ASTA Concept/Shutterstock)

to show others the insights about themselves that the tracker has gained, as well as by displaying their ability to make this data beautiful or easy to understand. For example, some runners use GPS tracking data to visualize their running routes in creative shapes. This type of data display is meant to attract attention and aesthetic curiosity and is an integral part of the Quantified Self movement (Lupton, 2017).

The genetic testing rabbit hole

One of the most hyped developments in the entire self-tracking industry in recent years is the Direct-To-Consumer (DTC) Genomics. Such services provide consumers a wide range of personal genetic information, including family DNA history, personal traits such as alcohol flush reaction to venous thromboembolism, to the more advanced detection of certain diseases or conditions—all based on the affordable and convenient home-collected blood or saliva samples. One company that has been rather successful in this area is 23andMe. Unlike its competitors, 23andMe is among the earliest to notice the potential of using social networks for marketing its products and services. Users can share their genetic results with others through the company's site or smartphone app and build social networks based on their genetic similarities.

However, not all discussions and social interactions that happen around people's genetic information are romantic (DNA matchmaking) or heartwarming (finding long-lost family members). On some corners of the Internet, if one digs further, there will be troubling signs. In one study, for example, Mittos and colleagues (2020) investigated how alt- and far-right groups talk about genetic testing on platforms such as Reddit and 4Chan. The researchers collected over 1.3 million comments related to genetic testing on the two platforms and subjected the data to computerized algorithmic analysis. Results show that discussions on genetic testing topics drew heavy traffic on both sites. On 4chan's politically incorrect board (/pol/), users posted large numbers of images with alt-right personalities (i.e., Carl Benjamin and Richard Spencer) and antisemitic memes. While on Reddit, despite the broader range of perspectives on the topic, conversations that contain racist, hateful, and misogynistic subject are not hard to find in some subreddits.

Have you encountered any discussions about genetic testing on social media platforms? If so, what kinds? In addition to Reddit and 4chan, have you observed similar discussions on other platforms? Do you believe that both social media and genetic testing companies have a responsibility to address and regulate such discussions?

A few issues generic to all forms of self-tracking are worth mentioning. First, the enormous wealth of data generated via self-tracking devices, be it blood pressure, heart rate, sleeping schedule, menstrual cycles, running route, and various biometrics, prompted the question regarding health-data ownership. Needless to say, users' health metrics are extremely valuable to tech companies, healthcare providers, and advertisers. As such, the right to determine who has access to this data and what to do with them is a constant battle in the age of quantified self. Stakeholders involved in all levels of the self-tracking industry must work closely to protect data privacy and security.

Second, while self-tracking experiments offer individuals a platform for self-discovery by illuminating unique aspects of the self and fostering personal agency (Kristensen & Ruckenstein, 2018), excessive reliance on health-tracking technology may result in addictive behavior (Motyl, 2020). In one study, for instance, researchers conducted interviews with 14 users who utilize Instagram to track their exercise habits over a period of 9 months (Kent, 2020). Participants initially experienced feelings of gratification, yet over time, many confessed to having addictive tendencies toward both Instagram and self-tracking technology due to the sensation of being under community scrutiny and the desire to maintain an ideal image.

To overcome its additive potential, users might consider periodic usage of self-tracking (incorporating periods of non-use or lapses), rather than continuous, frequent tracking (Gorm & Shklovski, 2019). However, the more significant concern lies beyond the addictive or compulsive use of tracking itself: as technology and social media platforms offer an increasing array of self-tracking options and produce a massive amount of data throughout an individual's lifetime, one can easily become overwhelmed in trying to make sense of it all.

Finally, some scholars have criticized that the current hype in the quantified-self movement resembles the notion of *knowing capitalism*, in which the attempt at knowing and analyzing our health data through various devices and apps becomes nothing but a way of sustaining the technology market (Ruckenstein & Pantzar, 2017). And this push to rethink life in a data-driven manner, or dataism, has the tendency to obfuscate some of the most critical epistemological inquires. For instance, the self-tracking devices sit on a fine line between luxury goods and basic needs: while some users think about matching his or her Apple watch with certain designer watch bands, others struggle to get the most basic pulse oximeter in the early days of the pandemic. And there is also the issue of how specific self-tracking devices and the data they generate contribute to new forms of sociality and politics. In some parts of the world, technologies such as health QR codes and contact-tracing apps has been misused (intentionally or unintentionally) by government and commercial entities for reasons such as mass surveillance, social control, workplace/employee monitoring, or, in some cases, outright profit. Therefore, as a society, we must collectively reckon the meaning and implications of these tracking systems closely and continuously.

Keywords: Health information seeking; uncertainty reduction; emotional support; health destigmatization; telehealth self-tracking; quantified self; fitspiration; knowing capitalism.

References

Bestsennyy, O., Gilbert, G., Harris, A., & Rost, J. (2021). *Telehealth: A quarter-trillion-dollar post-COVID-19 reality?* www.mckinsey.com/industries/healthcare-systems-and-services/our-insights/telehealth-a-quarter-trillion-dollar-post-covid-19-reality

Bryan, M. A., Evans, Y., Morishita, C., Midamba, N., & Moreno, M. (2020). Parental perceptions of the internet and social media as a source of pediatric health information. *Academic Pediatrics*, 20(1), 31–38. https://doi.org/10.1016/j.acap.2019.09.009

Burke, T. J., & Rains, S. A. (2019). The paradoxical outcomes of observing others' exercise behavior on social network sites: Friends' exercise posts, exercise attitudes, and weight concern. *Health Communication*, 34(4), 475–483. https://doi.org/10.1080/10410236.2018.1428404

De Las Heras-Pedrosa, C., Rando-Cueto, D., Jambrino-Maldonado, C., & Paniagua-Rojano, F. J. (2020). Analysis and study of hospital communication via social media from the patient perspective. *Cogent Social Sciences*, 6(1), 1718578. https://doi.org/10.1080/23311886.2020.1718578

Giedinghagen, A. (2022). The tic in TikTok and (where) all systems go: Mass social media-induced illness and Munchausen's by internet as explanatory models for social media associated abnormal illness behavior. *Clinical Child Psychology and Psychiatry*. https://doi.org/10.1177/13591045221098522

Gorm, N., & Shklovski, I. (2019). Episodic use: Practices of care in self-tracking. *New Media & Society*, 21(11–12), 2505–2521. https://doi.org/10.1177/1461444819851239

Gregg, M. (2015). Inside the data spectacle. *Television & New Media*, 16, 37–51. https://doi.org/10.1177/1527476414547774

Jargon, J. (2021). Teen girls are developing tics. Doctors say TikTok could be a factor. *The Wall Street Journal*. https://www.wsj.com/articles/teen-girls-are-developing-tics-doctors-say-tiktok-could-be-a-factor-11634389201

Kent, R. (2020). Self-tracking health over time: From the use of Instagram to perform optimal health to the protective shield of the digital detox. *Social Media + Society*, 6(3). https://doi.org/10.1177/2056305120940694

Koonin, L. M., Hoots, B., Tsang, C. A., Leroy, Z., Farris, K., Jolly, B., Antall, P., McCabe, B., Zelis, C. B., Tong, I., & Harris, A. M. (2020). Trends in the use of telehealth during the emergence of the COVID-19 pandemic—United States, January–March 2020. *Morbidity and Mortality Weekly Report*, 69(43), 1595. http://dx.doi.org/10.15585/mmwr.mm6943a3

Kristensen, D. B., & Ruckenstein, M. (2018). Co-evolving with self-tracking technologies. *New Media & Society*, 20(10), 3624–3640. https://doi.org/10.1177/1461444818755650

Leonard, B. (2022). Health data breaches swell in 2021 amid hacking surge, POLITICO analysis finds. *politico.com*. www.politico.com/news/2022/03/23/health-data-breaches-2021-hacking-surge-politico-00019283

Libert, T. (2015). Privacy implications of health information seeking on the web. *Communications of the ACM*, 58(3), 68–77. https://doi.org/10.1145/2658983

Lin, W. Y., Zhang, X., Song, H., & Omori, K. (2016). Health information seeking in the web 2.0 age: Trust in social media, uncertainty reduction, and self-disclosure. *Computers in Human Behavior*, *56*, 289–294. https://doi.org/10.1016/j.chb.2015.11.055

Lupton, D. (2017). Lively data, social fitness and biovalue: The intersections of health and fitness self-tracking and social media. *The SAGE handbook of social media* (pp. 562–578). Sage.

Malani. (2018). *Is social media changing the doctor-patient relationship?* https://ihpi.umich.edu/news/social-media-changing-doctor-patient-relationship

Mittos, A., Zannettou, S., Blackburn, J., & De Cristofaro, E. (2020). "And we will fight for our race!" A measurement study of genetic testing conversations on Reddit and 4chan. In *Proceedings of the international AAAI conference on web and social media* (Vol. 14, pp. 452–463).

Moorhead, S. A., Hazlett, D. E., Harrison, L., Carroll, J. K., Irwin, A., & Hoving, C. (2013). A new dimension of health care: Systematic review of the uses, benefits, and limitations of social media for health communication. *Journal of Medical Internet Research*, *15*(4), e1933. https://doi.org/10.2196/jmir.1933

Motyl, K. (2020). Compulsive self-tracking: When quantifying the body becomes an addiction. In Reichardt & Schober (Eds.), *Laboring bodies and the quantified self* (pp. 167–188). transcript Verlag. https://doi.org/10.1515/9783839449219-009

Myrick, J. G., Holton, A. E., Himelboim, I., & Love, B. (2016). # Stupidcancer: Exploring a typology of social support and the role of emotional expression in a social media community. *Health Communication*, *31*(5), 596–605. https://doi.org/10.1080/10410236.2014.981664

Neff, G., & Nafus, D. (2016). *Self-tracking*. MIT Press. https://doi.org/10.7551/mitpress/10421.001.0001

Obermeyer, Z., Powers, B., Vogeli, C., & Mullainathan, S. (2019). Dissecting racial bias in an algorithm used to manage the health of populations. *Science*, *366*(6464), 447–453. https://doi.org/10.1126/science.aax2342

Richter, J. P., & Kazley, A. S. (2020). Social media: How hospital facebook activity may influence patient satisfaction. *Health Marketing Quarterly*, *37*(1), 1–9.

Robinson, L., Prichard, I., Nikolaidis, A., Drummond, C., Drummond, M., & Tiggemann, M. (2017). Idealised media images: The effect of fitspiration imagery on body satisfaction and exercise behaviour. *Body Image*, *22*, 65–71. https://doi.org/10.1016/j.bodyim.2017.06.001

Rousidis, D., Koukaras, P., & Tjortjis, C. (2020). Social media prediction: A literature review. *Multimedia Tools and Applications*, *79*(9), 6279–6311. https://doi.org/10.1007/s11042-019-08291-9

Ruckenstein, M., & Pantzar, M. (2017). Beyond the quantified self: Thematic exploration of a dataistic paradigm. *New Media & Society*, *19*(3), 401–418. https://doi.org/10.1177/1461444815609081

Tennant, B., Stellefson, M., Dodd, V., Chaney, B., Chaney, D., Paige, S., & Alber, J. (2015). eHealth literacy and web 2.0 health information seeking behaviors among baby boomers and older adults. *Journal of Medical Internet Research*, *17*(3), e3992. https://doi.org/10.2196/jmir.3992

Tiggemann, M., Martins, Y., & Kirkbride, A. (2007). Oh to be lean and muscular: Body image ideals in gay and heterosexual men. *Psychology of Men & Masculinity*, *8*(1), 15. https://doi.org/10.1037/1524-9220.8.1.15

Tiggemann, M., & Zaccardo, M. (2018). Strong is the new skinny: A content analysis of #fitspiration images on Instagram. *Journal of Health Psychology*, *23*(8), 1003–1011. https://doi.org/10.1177/1359105316639436

Ventola, C. L. (2014). Social media and health care professionals: Benefits, risks, and best practices. *Pharmacy and Therapeutics*, *39*(7), 491.

Wang, A. [@alexandr_wang]. (2022, June 2). *Had wisdom teeth surgery last Thursday, and experiencing a bunch of pain tonight on one side of my jaw* [Tweet]. Twitter. https://twitter.com/alexandr_wang/status/1532276085320130565?lang=en

Zhao, W. (2022). The influence of media exposure on young women's intention to undergo cosmetic surgery: A third person perspective. *Atlantic Journal of Communication*, *30*(2), 146–158. https://doi.org/10.1080/15456870.2020.1856106

Part III

9 Social media for news and information sharing

In one of the best literary works, *A Tale of Two Cities*, British writer Charles Dickens described the zeitgeist of the 18th century Europe:

> It was the best of times, it was the worst of times, it was the age of wisdom, it was the age of foolishness, it was the epoch of belief, it was the epoch of incredulity, it was the season of Light, it was the season of Darkness, it was the spring of hope, it was the winter of despair, we had everything before us, we had nothing before us.

From an informational standpoint, this also applies to the world we live in nowadays. It is the best time, for we no longer rely on just a single source of information, such as *Acta Diurna*, the daily government gazette carved on stone presented in public places in 1st Century Rome. Instead, we now have all the news fingertips away, available 24/7, free of charge (sort of). Yet, it is the worst time, for, with all we know, a considerable portion of the population still believes in flat-earth theory, that a former leading candidate for the White house sexually malaises children for years in a Pizza joint, and that 5G towers are responsible for pandemics. To better understand this contrast, this chapter seeks to explore the role of social media in news consumption. Along the way, we will also examine questions such as how does news spread online? Who is likely to fall prey to various misinformation and disinformation? And why, beyond all its pejorative existence, social media may contribute to an open and functional society.

> **Exercise**
>
> As you read this chapter, the latest statistics on Americans' news consumption across social media may have been made available by Pew Research Center and other professional polling organizations. Try

DOI: 10.4324/9781003351962-12

> locating these reports online to examine the trends. Which are the leading platforms for news consumption in America? What platforms have seen the most growth in the percentage of users who regularly get news through them, compared to previous years? Additionally, pay close attention to the demographic profiles of regular social media news consumers across platforms. Which platforms tend to have disproportionately more female or male news users? Are there any platforms that seem more attractive to minority news users? Investigate the reasons behind this. Finally, if you work for one of these polling agencies, what other platforms would you include in this survey? Are there any other questions you would like to ask regarding people's social media news consumption?

Social media for news use

Let's start with some basic facts. According to the Pew Research center (2021), 67% percent of US adults reported receiving at least some news from social media. Among this population, roughly 31% indicated that they get news regularly on Facebook, far more than those who regularly use YouTube (22%) and Twitter (13%) for news. However, a different picture emerges when examining the proportion of each social media site users who regularly get news there. For example, despite Twitter being used by 23% of US adults, more than half of those users (55%) regularly get news on the site. In comparison, less than one-third of YouTube users regularly visit the platform for news consumption. Thus, on the surface level, some sites seem to possess more robust (or more "newsworthy") news attributes than others, indicating that the distinctive affordances of the platforms attract users for different purposes.

When it comes to news consumption on social media, users are not mere passive consumers but also active sharers. According to research by Kümpel et al. (2015), there are three broad categories of motives that drive people to share news on social media: self-serving, altruistic, and social motives. For instance, individuals may share news for self-serving reasons, such as gaining reputation or status, by sharing a post that aligns with their political views and attracts attention from their followers. Alternatively, individuals may share news for altruistic purposes, such as sharing an article about the latest medical breakthrough to inform and help their friends and family stay healthy. Lastly, social motives involve using news as a tool for social interaction, such as sharing a news article about a popular TV show to initiate a discussion with friends who share the same interest.

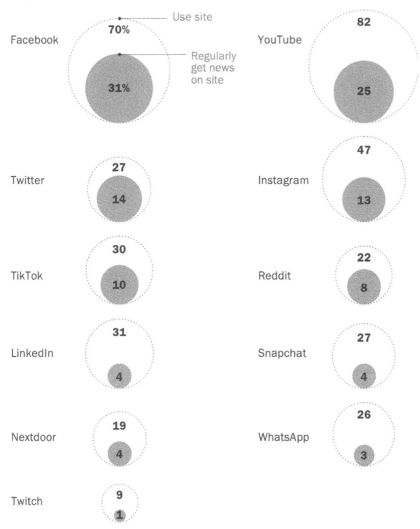

Figure 9.1 News consumption and use by social media sites
(*Source*: Pew Research Center)

In addition to these motivations, social network factors also play a significant role in news sharing (Ji, 2017). Tie strength, or the degree of closeness in a relationship, can influence news sharing, as individuals are more likely to share news with close friends and family members. The size and quality of one's social network, as well as homophily, can also affect news sharing. Individuals who view themselves as experts in a particular area, consume news frequently, and follow news organizations are more likely to share relevant information. Finally, individuals who value social connections and conversation and have a positive attitude toward sharing are more likely to engage in news sharing.

Incidental exposure to news

When you think about the latest news headlines, where did you first come across them? Chances are, some of them came to your attention somewhat serendipitously, be it the news soundbite from NPR on your way to school, the breaking news segment from a national TV network shown on the airport, or brief conversation with your roommates. These accidental exposures to news, as opposed to people actively searching for it, are referred to as *incidental exposure*.

In today's world, incidental exposure to news is likely to happen as you scroll through your Instagram or Twitter feed or auto-play the next YouTube video or TikTok clip, due to the ubiquity of news shared via social media. The phenomenon of incidental exposure reveals that there are some in our society who are not so interested in serious news, such as politics, yet they remain informed about major events.

For many years, political communication scholars have investigated whether unintentional forms of political information consumption led to any positive civic outcomes, such as voting, political discussion, and joining campaigns. However, research has provided inconsistent findings, with some studies suggesting positive effects of incidental exposure on knowledge or participation, while others show no effect or conditional relationships. For instance, recent meta-analysis by Nanz and Matthes (2022) revealed strong relationships between incidental exposure and news use, expressive engagement, and political discussion and slightly smaller relationships for political participation and knowledge. However, experimental studies that mimic incidental exposure to news reported only a small but significant effect on one's political knowledge. Some scholars have suggested that highly interested individuals may be entangled in a positive feedback loop with the algorithmic system that flushes more relevant incidental exposure

> content into their newsfeed. In contrast, individuals who mainly stay in first-level incidental exposure may have fewer opportunities for incidental exposure due to the lack of positive feedback for the algorithm. On that basis, Thorson (2020) argued that perhaps focusing on individuals' *attraction* to news makes more sense than focusing on the accidental nature of incidental exposure.

Echo chambers and filter bubbles

A major issue oft-invoked in the context of social media news sharing and consumption is the phenomenon of echo chambers (Sunstein, 2009). According to Jamieson and Cappella (2008), echo chambers are "a bounded, enclosed media space that has the potential to both magnify the messages delivered within it and insulate them from rebuttal" (p. 76). Essentially, echo chambers refer to a message environment in which the opinions, political leanings, or beliefs of users about a topic get reinforced due to repeated interactions with peers or sources having similar tendencies and attitudes. Researchers uncovered the presence of homophilic clusters of users on four platforms (Gab, Facebook, Reddit, and Twitter) in a study of over 100 million pieces of content on controversial topics, indicating the presence of echo chambers on social media (Cinelli et al., 2021). However, as we have discussed throughout the book, not all platforms are created equal—when comparing Facebook with Reddit, researchers found a higher level of segregation in terms of news consumption on Facebook. While on Reddit, even users who are leaning toward extreme views tend to interact with the majority. The researchers attributed such difference partially to the fact that Reddit's feed algorithm is more tweakable by its users than that of Facebook and Twitter.

Predictably, the concern over algorithms' impact on how news and information are distributed online has prompted activists and entrepreneurs such as Eli Pariser to coin the term "filter bubbles" (Pariser, 2011), which denotes algorithms inadvertently amplifying ideological segregation by automatically recommending content an individual is likely to agree with.

Although the concept of echo chambers and filter bubbles may seem intuitive, some have questioned their actual prevalence on social media. A growing number of studies suggest that, despite people's self-selection and confirmation bias, echo chambers are less widespread than commonly assumed. Furthermore, there is minimal empirical evidence directly supporting the notion that algorithms cause political polarization. In fact, recent studies, conducted across various election cycles, have discovered that algorithms have no significant impact on people's beliefs and voting intentions. (Guess, et al., 2023; Groshek & Koc-Michalska, 2017; Ross et al., 2022).

Fake news, misinformation, and disinformation

Despite the ongoing debate surrounding filter bubbles and echo chambers, few dispute the existence and threat of "fake news" on social media. However, depending on its usage, fake news has become a cliché for any information that people do not believe in. A less value-laden term, at least among journalists and academia, would be misinformation, which refers to "a category of claim for which there is substantial disagreement (or even consensus rejection) when judged for truth-value among the widest feasible range of observers" (Southwell et al., 2021, p. 3). To illustrate, consider the issue of climate change, for which over 99% of scientific papers agree on its existence and primary cause (i.e., human activities), yet it is still heavily contested by many deniers who argue that the problem is either an exaggeration or a downright hoax (Biddlestone & van der Linden, 2021). And, to date, one can easily find a wealth of unreliable discourse on social media that happily absolves our responsibility to contain climate change.

Typologies of fake news

Researchers had already been studying "fake news" for decades before certain politicians made the term a common language (and hence lost its communication value). Tandoc Jr. and colleagues (2017) examined 34 academic articles that used the term "fake news" between 2003 and 2017, aiming to decipher how fake news as an evolving phenomenon was interpreted by researchers. Their analysis yielded six types of fake news; namely, news satire, news parody, fabrication, manipulation, advertising, and propaganda.

A news satire is a mock news program that uses humor or exaggeration when reporting on current events. They use the same style as a regular news broadcast would, but they aim to deliver entertainment before information. However, what they present is still real news; the format is what is most considered to be "fake."

A news parody is similar to a satire in its humor and presentation format; however, parodies use non-factual information to create humor, sometimes even completely making up news stories to be comical. In satire and parody, both the author and the reader are "in" on the fact that it is "fake news."

News fabrication refers to articles that are completely false in terms of the information they present but are styled to look legitimate. The purpose of fabrication is to misinform, and in these situations, the reader is not always "in" on the joke, causing a possible spread of misinformation.

Photo manipulation is altering photos or videos to create a false narrative, making them look like something that they are not. Manipulation can also occur without altering the photo at all but instead associating it with a different and falsified context. (This type of fake news is increasingly common on the Internet with the advancement of the so-called deepfake AI-generated images, such as Pope Francis wearing a white puffer jacket or riding a motorcycle.)

Advertising is press releases that are published as news. Oftentimes, advertising campaigns will do whatever it takes—even if that includes exaggeration – to generate attention and revenue for their products and companies.

Finally, propaganda refers to news stories created by a political entity to influence the public's perceptions with the intent to benefit a public figure, government, or organization.

Now that you have mastered this typology, can you find an example of each type of fake news on your go-to social media platform?

Another term that often gets conflated with misinformation is the notion of disinformation—"the deliberate creation and sharing of information known to be false" (Wardle, 2017). By definition, disinformation is deceptive in its intent. For instance, in the early days of the Russian-Ukraine war, reports found that the Russian government deliberately engaged in disinformation campaigns on TikTok and Telegram, spreading a slew of anti-Ukrainian content (Klepper, 2022). In contrast, misinformation sharing may be less intentional, but is often motivated by a range of factors, such as individuals' confirmation bias, motivated reasoning, ideological predispositions, as well as the structure of online and offline information networks (Valenzuela et al., 2019).

Viral news

Average social media users are less likely to delineate the subtle conceptual differences between misinformation and disinformation; instead, they are more likely to be drawn toward popular and viral content. This raises the question: What makes certain news stories become viral? Recent research has provided some insight into the content-related and presentation-related factors that contribute to the success of news stories.

Content-related factors relate to the attributes of the news story itself, including its valence, interestingness, and the issues and topics discussed. On that, studies have shown that *positive* news stories that are informative, trustworthy, and interesting are more likely to be shared (Berger & Milkman, 2012; Ji et al., 2020). A prime example is the story of the "Ice Bucket Challenge" in 2014. The challenge, which involved people dumping a bucket of ice water on their heads and nominating others to do the same, was originally started as a way to raise awareness and funds for amyotrophic lateral sclerosis (ALS), also known as Lou Gehrig's disease. The positive and charitable nature of the challenge, combined with its novelty and the participation of high-profile celebrities, made it highly shareable and resulted in it becoming a viral sensation on social media.

In addition to content-related factors, presentation-related factors also affect the virality of news stories. Simply put, presentation-related factors pertain to the ways in which news stories are presented online. Consider the "dress" phenomenon that took the internet by storm in 2015. The story originated from a simple photo of a dress that some people perceived as blue and black, while others saw it as white and gold. The photo was shared widely on social media, and the controversy surrounding the color of the dress created a sense of intrigue and discussion that drove its virality. The visual nature of the photo and the debate it sparked, along with the ways in which it was presented and shared online, all contributed to its success in going viral. However, many social media companies have gone beyond simply allowing users to manipulate the visual presentation of news stories; they have invented new ways and indices, such as article ratings, comments, view counts, and the ranking or placement of an article on a given page, to signal the prominence of each story (Li & Sakamoto, 2014). As a result, some stories have gone viral simply because they caught the eyes of a small group of readers in the first place.

News fact-checking

In the wake of the fake news epidemic and its negative impact on civil societies, countries across the globe have started to explore effective ways to structurally combat misinformation and maintain the integrity of their information ecosystem. One particular measure that has been regarded with high hopes is the century-old practice in the journalism industry known as news fact-checking.

Broadly, news fact-checking refers to verifying the accuracy of news reports and information, with the ultimate goal of the public having access to accurate and reliable information (Allcott & Gentzkow, 2017). In practice, news fact-checking involves checking sources, verifying quotes and statements, and determining the context in which information is presented (Gentzkow et al., 2016). While many legacy news agencies such as the Associated Press, Reuters, BBC, and Deutsche Welle have their own dedicated internal fact-checking teams, recent years have seen specialized fact-checking organizations being established by academics, news practitioners, and subject experts (Graves & Amazeen, 2019). Notable examples include PolitiFact, FactCheck.org, and Snopes.

So, does news fact-checking actually work? Fortunately, studies have shown that, by and large, fact-checking can be effective in helping to fight fake news, with average news users becoming more attuned to accurate information after a single exposure to a fact-checking message (Walter et al., 2020). What is more, news users seem willing to engage with (by means of sharing and commenting) fact-checking messages that incorporate statistics and official reports to hence their source transparency. Similarly, contextual information, such as why the claim under scrutiny occurred and why fact-checking the claim was necessary, also contributed to readers' engagement with fact-checking messages (Kim et al., 2021).

However, it is a common experience that most users don't read the full news story on social media before sharing or commenting. Instead, they glimpse news headlines and decide whether to engage the story or move on to the next thing on their feed. To alert users about certain problematic headlines, some social media platforms (i.e., Facebook, Twitter, and Google) practiced flagging or tagging news stories that were deemed problematic by their partnered fact-checkers. Emerging online experiments that subject users to headlines marked with and without warning messages found that such practices can significantly reduce users' belief about the accuracy of the news headline. However, warning messages appear ineffective in deterring users from *sharing* false headlines (Clayton et al., 2020). It is partly for this reason that Facebook abolished flagging false news in late 2017 and re-oriented toward reducing the distribution of false information and providing more contextual information around the disputed news content (Andersen & Søe, 2020).

At this point, we ought to recognize that news fact-checking is a helpful tool for addressing issues about social media fake news and misinformation. But it is far from a magic bullet. A host of other factors, such as source expertise/credibility, the partisanship of the users and news outlet, users' news/information literacy, media skepticism, as well as the format and tone of the fact-checking message, all affect users' acceptance toward the debunked message (Schmid & Betsch, 2019; Nekmat, 2020; Young et al., 2018). Therefore, social media platforms and government regulatory bodies must consider other approaches rather than relying on news fact-checking as the sole counteractive measure (Bavel et al., 2020).

Conspiracy theories

But what about conspiracy theories? Are social media responsible for spreading dubious ideas about electoral fraud, COVID-19 vaccine safety, and satanic pedophiles controlling the government? Ask average Americans; nearly 75% of respondents indicated that social media (and the Internet in general) is the primary venue whereby conspiracy theories spread (Quinnipiac University Polling Institute, 2021). Moreover, many studies seem to support at least some correlations between social media use and conspiratorial beliefs (e.g., Allington & Dhavan, 2020; Bridgman et al., 2020). Interestingly though, national surveys also show that the percentage of Americans who endorse specific conspiracy theories is not rising; even COVID-19 conspiracies and support for QAnon remained largely stable/unchanged during the period when related information was widely available on social media (Enders et al., 2020; Romer & Jamieson, 2020). So, how do we make sense of this seemingly contradictory status quo?

Fortunately, the latest research can provide at least some clues. In a survey research with nationally representative samples, Enders and colleagues (2021) asked over 3,000 Americans about their media use and beliefs in a list of conspiratorial statements (e.g., "There is a 'deep state' embedded in the government that operates in secret and without oversight"). Consistent

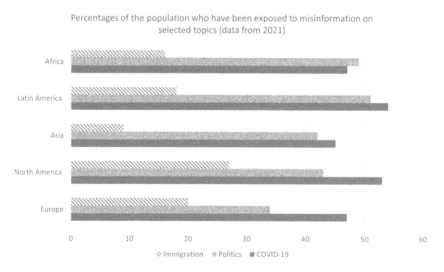

Figure 9.2 Percentages of the population who have been exposed to misinformation on selected topics

(*Source*: Hölig et al., 2022)

with other similar studies, it was uncovered that more social media usage (across platforms) as a whole positively correlated with conspiracy beliefs. However, further analysis showed that this association was conditioned by the notion of conspiracy thinking—a predisposition or tendency to impose conspiratorial narratives on salient affairs (Enders et al., 2021). In other words, the connection between social media use and conspiracy beliefs tends to be stronger among people who are geared toward conspiracy thinking. This suggests that the prevailing narrative that social media widely spreads conspiracy theories and misinformation is perhaps ill-fitted, and conspiracy theories tend to adhere to minds that are already hospitable to them. Indeed, robust evidence from 26 countries has shown that conspiracy thinking is associated with extreme political ideologies; more so among those with far-right beliefs than far-left beliefs (Imhoff et al., 2022). As Alice Marwick, Co-Director of The Center for Information, Technology, and Public Life at University of North Carolina at Chapel Hill puts it in an interview (Cohn & Kelley, 2023, 8:00):

> One of the reasons why QAnon (and a few other conspiracies) has been so successful, is that it picks up a lot of other pre-circulating conspiracy theories, it mixes them with antisemitism . . . homophobia and transphobia, and creates this hideous concoction . . . that reinforces there pre-existing beliefs.

Policy and legal challenges regarding social media news consumption

The popularity of social media for news consumption also poses challenges to policy-makers and business owners, leading to a number of legal and regulatory concerns. For one thing, scholars and news publishers have long argued that tech giants and social media platforms profit from the news industry and eventually pose threats to the health of democracy, citing that sites and services such as Google and Facebook gain traction (hence adverting revenues) by circulating news content on its sites yet ignoring the cost to their original publishers. While some international publishers such as France's Agency France-Presse, Germany's Deutsche Presse-Agentur, and Britain's Press Association rallied together calling for social media companies to pay for news content used on its sites, other elite publishers, including *The New York Times, The Guardian*, and the *Hearst Corporation*, decided to opt-out of certain news aggregation services (e.g., Apple news, Facebook instant news). Corporations responded differently: some decided to pay publishers, whereas others (Facebook) escalated the stalemates by restricting news sharing in countries that demand payment (Isacca, 2021).

> **Taking over the town square**
>
> On April 26, 2022, the founder of Tesla and Space X, Elon Musk, reached an agreement with Twitter, allowing the business tycoon to buy over the popular micro-blogging platform for a sum of $44 billion. Public opinion exploded, with many voicing concerns about Twitter's potential efforts to regulate online speech. In a TED interview shortly after the deal was announced to the public, Musk explained that the platform "has become kind of the de facto town square, so it's just really important that people have both the reality and the perception that they are able to speak freely within the bounds of the law."
>
> In many ways, the concern over Musk's takeover of Twitter is not unfounded. After all, the controversial businessman has advocated for less content moderation on Twitter for years against the backdrop of Twitter granting users the ability to mute, block, or report content/users related to hate speech, harassment, and medical misinformation. Sometimes, accounts may be removed or suspended by the platform for speech deemed harmful to society. One prominent case was Twitter's decision to permanently ban former President Donald Trump after January 6, 2021. Many people, particularly those on the political left, believe content moderation is an effective way to limit the impact of harmful information, whereas people on the right condemned Twitter for conducting "censorship."
>
> In reality, the mixed views on content moderation among the public reflect the complex feelings that users are experiencing. Research has suggested that most content moderation systems, as currently implemented, remove content at large scales yet do little to educate users about where they went wrong. Additionally, the absence of an effective appeals system has led users to develop folk theories about how platform moderation works and blame human intervention as the primary cause of removing their content (Myers West, 2018). In previous chapters, we discussed the potential adverse effects of social media content on some vulnerable populations and general people in specific settings (e.g., health and dating). Therefore, the debate about social media content moderation is perhaps less about "whether or not" but more about "how."

Debate exists on whether social media companies have the right to censor or remove speech on their platforms. Supporters argue that social media has grown too big and too powerful, making it a hotbed for misinformation and disinformation, particularly hate speech. For example, a *Times* report uncovered that Facebook's operation (or lack of engagement) in Myanmar unwittingly amplified rumors and disinformation about the

Rohingya people, resulting in a bloody genocide in 2017. Therefore, social media companies must consider some form of content moderation as an essential supplement for their services. In the early days of YouTube, the platform was quick to notice that user-generated content, such as pornography and extreme violence, jeopardized the core of their business. As such, they formed one of the earliest specialized content moderation team in the Silicon Valley (fittingly named as SQUAD: Safety, Quality, and User Advocacy) to filter those undesirable videos (Bergen, 2022). On the other hand, opponents argue that companies have private ownership over that content and people's right to the First Amendment should be respected.

In addition to social media companies' efforts, government bodies in parts of the world are responding to growing concerns over social media misinformation. Notably, in July 2022, the European Parliament passed a landmark legislation named the Digital Services Act (DSA), mandating social media companies to do more to moderate harmful content. Under this new legislation, companies are likely to face hefty fines for DSA breaches, making content moderation almost a requirement in EU countries (Chee, 2022). However, some controversial politicians or entrepreneurs in the US have started their own platforms (e.g., Truth Social) or privatized existing ones (e.g., Elon Musk's acquisition of Twitter) to create safe havens for all kinds of speech.

Ultimately, the constant push and pull among companies, citizens, and governments will affect the collective views of a society about content moderation and free speech. However, in the campaign against misinformation, it is certain that no quick fix, single policy, or technological intervention strategy alone will work for all instances.

Keywords: Incidental exposure to news; social media news-consumption; news sharing; news fact-checking; echo chambers; filter bubbles; fake news; disinformation; misinformation; conspiracy theories; content moderation; Digital Services Act.

References

Allcott, H., & Gentzkow, M. (2017). Social media and fake news in the 2016 election. *Journal of Economic Perspectives*, *31*(2), 211–236.

Allington, D., & Dhavan, N. (2020). The relationship between conspiracy beliefs and compliance with public health guidance with regard to COVID-19. *The Center for Countering Digital Hate*. https://kclpure.kcl.ac.uk/portal/files/127 048 253/Allington_and_Dhavan_2020.pdf.

Andersen, J., & Søe, S. O. (2020). Communicative actions we live by: The problem with fact-checking, tagging or flagging fake news–the case of Facebook. *European Journal of Communication*, *35*(2), 126–139. https://doi.org/10.1177/0267323119894489

Bandari, R., Asur, S., & Huberman, B. S. (2012). *The pulse of news in social media: Forecasting popularity* [Paper presentation]. Proceedings of the Sixth

International AAAI Conference on Weblogs and Social Media (Vol. 1202, pp. 26–33), (Arxiv preprint arXiv), Dublin, 4–7 June.

Bavel, J. J. V., Baicker, K., Boggio, P. S., Capraro, V., Cichocka, A., Cikara, M., Crockett, M. J., Crum, A. J., Douglas, K. M., Druckman, J. N., Drury, J., Dube, O., Ellemers, N., Finkel, E. J., Fowler, J. H., Gelfand, M., Han, S., Alexander Haslam, S., Jetten, J., . . . Willer, R. (2020). Using social and behavioural science to support COVID-19 pandemic response. *Nature Human Behaviour*, 4(5), 460–471. https://doi.org/10.1038/s41562-020-0884-z

Bergen, M. (2022). *Like, comment, subscribe: Inside YouTube's chaotic rise to world domination*. Random House Large Print.

Berger, J., & Milkman, K. L. (2012). What makes online content viral? *Journal of Marketing Research*, 49, 192–205. https://doi.org/10.1509/jmr.10.0353

Biddlestone, M., & van der Linden, S. (2021). Climate change misinformation fools too many people – but there are ways to combat it. *theconversation.com*. https://theconversation.com/climate-change-misinformation-fools-too-many-people-but-there-are-ways-to-combat-it-170658

Bridgman, A., Merkley, E., Loewen, P. J., Owen, T., Ruths, D., Teichmann, L., & Zhilin, O. (2020). The causes and consequences of COVID-19 misperceptions: Understanding the role of news and social media. *Harvard Kennedy School Misinformation Review*, 1(3). https://doi.org/10.37016/mr-2020-028

Chee, F. Y. (2022). EU lawmakers pass landmark tech rules, but enforcement a worry. *reuters.com*. www.reuters.com/technology/eu-lawmakers-pass-landmark-tech-rules-enforcement-worry-2022-07-05/

Cinelli, M., Morales, G. D. F., Galeazzi, A., Quattrociocchi, W., & Starnini, M. (2021). The echo chamber effect on social media. *Proceedings of the National Academy of Sciences*, 118(9). https://doi.org/10.1073/pnas.2023301118

Clayton, K., Blair, S., Busam, J. A., Forstner, S., Glance, J., Green, G., Kawata, A., Kovvuri, A., Martin, J., Morgan, E., Sandhu, M., Sang, R., Scholz-Bright, R., Welch, A. T., Wolff, A. G., Zhou, A., & Nyhan, B. (2020). Real solutions for fake news? Measuring the effectiveness of general warnings and fact-check tags in reducing belief in false stories on social media. *Political Behavior*, 42(4), 1073–1095. https://doi.org/10.1007/s11109-019-09533-0

Cohn, C., & Kelley, J. (2023, March 21). So you think you're a critical thinker. In *How to fix the internet*. https://podcasts.apple.com/us/podcast/so-you-think-youre-a-critical-thinker/id1539719568?i=1000605122219

Enders, A. M., Uscinski, J. E., Klofstad, C., & Stoler, J. (2020). The different forms of COVID-19 misinformation and their consequences. *The Harvard Kennedy School Misinformation Review*. https://doi.org/10.37016/mr-2020-48

Enders, A. M., Uscinski, J. E., Seelig, M. I., Klofstad, C. A., Wuchty, S., Funchion, J. R., Murthi, M. N., Premaratne, K., & Stoler, J. (2021). The relationship between social media use and beliefs in conspiracy theories and misinformation. *Political Behavior*, 1–24. https://doi.org/10.1007/s11109-021-09734-6

Gentzkow, M., Shapiro, J., & Taddy, M. (2016). *Measuring polarization in high-dimensional data: Method and application to congressional speech* (No. id: 11114).

Graves, L., & Amazeen, M. A. (2019). Fact-checking as idea and practice in journalism. In *Oxford research encyclopedia of communication.* https://doi.org/10.1093/acrefore/9780190228613.013.808

Groshek, J., & Koc-Michalska, K. (2017). Helping populism win? Social media use, filter bubbles, and support for populist presidential candidates in the 2016 US election campaign. *Information, Communication & Society, 20*(9), 1389–1407. https://doi.org/10.1080/1369118x.2017.1329334

Guess, A., Malhotra, N., Pan, J. Barbera, P. . . . (2023) How do social media feed algorithms affect attitudes and behavior in an election campaign? Science. 381 (6656). pp. 398–404 https://doi.org/10.1126/science.abp9364

Hölig, S., Behre, J., & Schulz, W. (2022). *Reuters institute digital news report 2022: Ergebnisse für Deutschland* https://leibniz-hbi.de/uploads/media/Publikationen/cms/media/k3u8e8z_AP63_RIDNR22_Deutschland.pdf.

Imhoff, R., Zimmer, F., Klein, O., António, J. H., Babinska, M., Bangerter, A., Bilewicz, M., Blanuša, N., Bovan, K., Bužarovska, R., Cichocka, A., Delouvée, S., Douglas, K. M., Dyrendal, A., Etienne, T., Gjoneska, B., Graf, S., Gualda, E., Hirschberger, G., . . . Van Prooijen, J. W. (2022). Conspiracy mentality and political orientation across 26 countries. *Nature Human Behaviour,* 1–12. https://doi.org/10.1038/s41562-021-01258-7

Isacca, M. (2021). Facebook restricts the sharing of news in Australia as Google says it will pay some publishers. *New York Times.* www.nytimes.com/2021/02/17/technology/facebook-restricts-the-sharing-of-news-in-australia-as-google-says-it-will-pay-some-publishers.html

Jamieson, K. H., & Cappella, J. N. (2008). *Echo chamber: Rush Limbaugh and the conservative media establishment.* Oxford University Press. https://doi.org/10.1002/j.1538-165x.2009.tb01921.x

Ji, Q. (2017). Social media news use and political discussion: A focus on Chinese users' news reception and dissemination. *Electronic News, 11*(1), 3–19. https://doi.org/10.1177/1931243116672257

Ji, Q., Raney, A. A., Janicke-Bowles, S. H., Dale, K. R., Oliver, M. B., Reed, A., Seibert, J., & Raney, A. A. (2020). Spreading the good news: Analyzing socially shared inspirational news content. *Journalism & Mass Communication Quarterly, 96*(3), 872–893. https://doi.org/10.1177/1077699018813096

Kim, H. S., Suh, Y. J., Kim, E. M., Chong, E., Hong, H., Song, B., Ko, Y., & Choi, J. S. (2021). Fact-checking and audience engagement: A study of content analysis and audience behavioral data of fact-checking coverage from news media. *Digital Journalism,* 1–20. https://doi.org/10.1080/21670811.2021.2006073

Klepper, D. (2022). *TikTok is Russia's newest weapon in arsenal for anti-Ukraine propaganda.* www.usatoday.com/story/tech/news/2022/02/26/russia-ukraine-tiktok-disinformation/6951910001/

Kümpel, A. S., Karnowski, V., & Keyling, T. (2015). News sharing in social media: A review of current research on news sharing users, content, and networks. *Social Media + Society, 1*(2). https://doi.org/10.1177/2056305115610141

Li, H., & Sakamoto, Y. (2014). Social impacts in social media: An examination of perceived truthfulness and sharing of information. *Computers in Human Behavior, 41,* 278–287.

Myers West, S. (2018). Censored, suspended, shadowbanned: User interpretations of content moderation on social media platforms. *New Media & Society*, *20*(11), 4366–4383. https://doi.org/10.1177/1461444818773059

Nanz, A., & Matthes, J. (2022). Democratic consequences of incidental exposure to political information: A meta-analysis. *Journal of Communication*. https://doi.org/10.1093/joc/jqac008

Nekmat, E. (2020). Nudge effect of fact-check alerts: Source influence and media skepticism on sharing of news misinformation in social media. *Social Media + Society*, *6*(1). https://doi.org/10.1177/2056305119897322

Pariser, E. (2011). *The filter bubble: What the internet is hiding from you*. Viking.

Quinnipiac University Polling Institute. (2021). Quinnipiac university poll, question 27 [31118210.00026]. In *Roper center for public opinion research*. Quinnipiac University Polling Institute.

Romer, D., & Jamieson, K. H. (2020). Conspiracy theories as barriers to controlling the spread of COVID-19 in the US. *Social Science & Medicine*, *263*, 113356. https://doi.org/10.1016/j.socscimed.2020.113356

Ross Arguedas, A., Robertson, C., Fletcher, R., & Nielsen, R. (2022). *Echo chambers, filter bubbles, and polarisation: A literature review*. Reuters Institute for the Study of Journalism. https://reutersinstitute.politics.ox.ac.uk/echo-chambers-filter-bubbles-and-polarisation-literature-review

Schmid, P., & Betsch, C. (2019). Effective strategies for rebutting science denialism in public discussions. *Nature Human Behaviour*, *3*(9), 931–939. https://doi.org/10.1038/s41562-019-0632-4

Southwell, B. G., Thorson, E. A., & Sheble, L. (2021). Introduction: Misinformation among mass audiences as a focus for inquiry. In *Misinformation and mass audiences* (pp. 1–12). University of Texas Press. https://doi.org/10.7560/314555-002

Sunstein, C. R. (2009). *Going to extremes: How like minds unite and divide*. Oxford University Press.

Tandoc, E. C., Jr., Lim, Z. W., & Ling, R. (2018). Defining "fake news" a typology of scholarly definitions. *Digital Journalism*, *6*(2), 137–153. https://doi.org/10.1080/21670811.2017.1360143

Thorson, K. (2020). Attracting the news: Algorithms, platforms, and reframing incidental exposure. *Journalism*, *21*(8), 1067–1082. https://doi.org/10.1177/1464884920915352

Valenzuela, S., Halpern, D., Katz, J. E., & Miranda, J. P. (2019). The paradox of participation versus misinformation: Social media, political engagement, and the spread of misinformation. *Digital Journalism*, *7*(6), 802–823. https://doi.org/10.1080/21670811.2019.1623701

Walter, N., Cohen, J., Holbert, R. L., & Morag, Y. (2020). Fact-checking: A meta-analysis of what works and for whom. *Political Communication*, *37*(3), 350–375. https://doi.org/10.1080/10584609.2019.1668894

Wardle, C. (2017). *Fake news. It's complicated*. https://medium.com/1st-draft/fake-news-its-complicated-d0f773766c79

Young, D. G., Jamieson, K. H., Poulsen, S., & Goldring, A. (2018). Fact-checking effectiveness as a function of format and tone: Evaluating FactCheck.org and FlackCheck.org. *Journalism & Mass Communication Quarterly*, *95*(1), 49–75. https://doi.org/10.1177/1077699017710453

10 Social media for social movement and political campaign

At Sidi Bouzid, a rural town in the North African country of Tunisia, corruption was ordinary. With an unemployment rate as high as 30%, life was not easy for the residents in town either. To support his family and himself, 27-year-old Mohamed Bouazizi applied to join the army but was rejected. His subsequent job applications elsewhere were also futile. To survive, he decided to take on some debt and sell fruit at a roadside stand. But his small business attracted unwanted attention from the local police officers. Eventually, on December 17, 2010, a municipal official confiscated his wares and reportedly harassed and humiliated Bouazizi during the process. Enraged by such an unjust experience, Bouazizi later went to the governor's office to make complaints but was rejected by the governor. And, out of desperation, the young man finally set himself afire.

Despite being sent to the hospital, Bouazizi was not going to survive. His death on January 4, 2011, eventually brought together various groups dissatisfied with the existing system to begin the Tunisian Revolution, which resulted in a regime change, and quickly spread into the entire Arab world in the early 2010s. Today, experts and political commentators refer to this chain of incidences as the Arab Spring; and social media were regarded as the holy grail for those protests to bear fruits.

Defining social movement

Social movement (a.k.a. collective activism or organized activism) is a social process through which people get together and voice their opinions, critiques, and sometimes grievances. There are three key features that set social movements apart from numerous types of collective actions such as fandom and organized political parties: first, social movements are conflictual in nature and often have clearly articulated opponents. Examples include the suppressive governing regimes in the Arab Spring,

Figure 10.1 Egyptian activists protesting during the Arab Spring
(*Source*: MidoSemsem/Shutterstock)

the sexual harassment perpetrators in #Metoo, or the exploitive social/political beliefs in the Civil Rights Movement of the 60s. Second, social movements must also be supported by dense yet informal networks; for that is how individuals' voices and actions can be organized and channeled into powerful forces. The informal nature of networks also makes social movements distinctive from political parties or corporations, where those memberships tend to be much more rigid. Lastly, successful collective activism tends to develop and sustain a set of collective identities, as in the cases of environmental activism and the Black Lives Matter movement, where a collective identity was formed around the shared goal of environmental or racial justice.

Given these features of social movements, one could argue that some health-related collective actions, such as the breast cancer movement, the tobacco control movement, and the anti-drunk-driving movement, despite all being colloquially addressed as "movements," are not what we would strictly consider social movements because of the lack of clearly defined opponents, the involvement of government or corporate forces, and the lack of a well-defined collective identity among the participants.

Exercise

Companies like GameStop and AMC Theatres are not high-profile tech companies that frequently appear in news headlines. However, in January 2021, the market experienced a sudden increase in demand for these companies' stocks, resulting in a rapid rise in their stock values. Unfortunately, the short squeeze (a term often used to describe this phenomenon in the stock market) caused many Wall Street investment agencies that bet against these companies' stocks to lose billions of dollars. The cause? A group of devoted users from the social media platform Reddit called for buying those companies' stocks shortly after some Wall Street short-sellers announced that they expected to see a decrease in value in those stocks. Commentators dubbed this event "activism investment" (Lyons, 2021), citing its possible connection with the anti-Wall Street movement in 2011. What is your view? Do you think this event can be considered a new form of social-media-empowered movement?

While social movements have existed before the digital media era, it is widely acknowledged that modern communication technologies have significantly influenced the initiation, growth, and spread of activism. Bennett and Segerberg (2013) introduced the concept of "connective action," which captures how the *social* nature of social media forms the basis for a movement, and people's co-production and co-distribution of ideas generate a self-propelling participatory system. Through connective action, large publics can engage in conversations about vital topics, find their voices, and take action, both nationally and transnationally. Thus, while social media did not create social movements, it has certainly revolutionized the way people mobilize and organize themselves for political and social change.

How do social media support protests and movements

In a review of key research on technology and social movement, Cammaerts (2015) outlined two broad areas where social media support protests and movements: a) they serve as tools for internal organization and communication; and b) they serve as weapons or instruments of ideological attack. Next, we will take a closer look at the role of social media in both accounts.

First, as tools, social media platforms, particularly messaging apps and various social network sites, serve as powerful tools for organizing and participating in movements and protests. Their ubiquity, relatively low cost, and synchronicity enable members of a movement to communicate, recruit, and network effectively, as well as to enter and exit a movement freely. This is especially important for today's digital natives. For example, Belotti and colleagues (2022) conducted a six-month ethnographic study of the FridaysForFuture movement (FFF), a climate movement launched by activist Greta Thunberg in 2018. Through participant observation of FFF assemblies and digital ethnography on WhatsApp, the researchers found that FFF activists, primarily young people, use social media as a key battleground, leveraging various platforms to raise awareness, obtain information, recruit supporters, and coordinate offline actions. Interestingly, these activists appear to be skilled at using different platforms to reach different audiences. For instance, they may use Twitter to engage with institutions and the mass media, while turning to TikTok to reach younger audiences. In addition, they may combine marches and sit-ins with artistic performances and educational activities in schools to further their cause.

The second function of social media as a communication and organizational tool is that it enables movement to spread across physical and temporal borders, channeling populations that are otherwise separated by time and space. The spill-over effect observed during the Arab Spring, where protests and uprisings in one country inspire similar movements in other countries, is a prominent example of such functions. Aside from political movements (which tend to attract extensive media coverage), other types of social movements also need to connect with national and transnational publics by discovering and amplifying individuals' voices and experiences over an extended period of time. In a study of feminist activism on Twitter, Li and colleagues (2021) examined how users became members of the movement against sexual assault. To do so, they scrapped tweets using the hashtags #WhyIDidntReport and #MeToo and performed a series of qualitative thematic analyses. It was revealed that Twitter users effectively used these hashtags to share resources, engaged in conversations relating to the political or policy aspects of the issue, and took steps to promote social actions, helping increase the social awareness and volume surrounding the topic of sexual assault. In particular, users' self-disclosure of their sexual assault experiences, such as the influence of the perpetrators, their sense of helplessness, and their secondary victimization by the police served as a powerful testament to the significance of the matter.

Successful movements tend to entail strong offline actions. In that regard, social media's third function as an instrument is to facilitate mobilization and coordination within movements, enabling on-the-spot or in-real-time communicative practices. For example, Urman and colleagues (2021) looked at the popular messaging app Telegram and how it was used by protesters during the 2019 Anti-Extradition Bill movement in Hong Kong. The movement was born out of concern about a proposed bill that would have allowed the Hong Kong government to send fugitives to jurisdictions without existing bilateral extradition agreements with Hong Kong, including Mainland China, Taiwan, and Macau. Telegram emerged as a major communication tool during the movement due to its perceived devotion to privacy and anonymity. Using the snowball-based sampling technique (given that Telegram does not allow developers to search its content), the researchers collected data from 1,806 public Telegram channels in the Hong Kong Telegram ecosystem during the protest. Subsequent textual analysis showed that a large amount of communication that happened on Telegram was about activists distributing information about police presence during protests, their protest-related actions, as well as deliberations. This demonstrates the instrumental role played by Telegram in the organization of the 2019 protests in Hong Kong.

Social movements can be quite fluid in their goals and strategies. Hence, members of a movement often find themselves engaging in internal debate and decision-making in order to formulate a clear path forward. In that sense, social media is a space where online deliberation happens. And, as the deliberation evolves, members of the movement generate texts, slogans, memes, and other audio-video content which facilitate the dissemination of its ideas and spirits with frames and perspectives that better resonates with people. For example, research showed that, during the Occupy Wall Street movement, activists and commentators used the hashtag #OWS to debate the goals and strategies of the movement, resulting in a multitude of different opinions about the protest and eventually helping people to know, quite literally, where to go and how to behave (Gleason, 2013). Moreover, in the aforementioned study about the #FFF movement, activists discussed heavily about whether or not to use TikTok as a channel to spread information and raise awareness about the climate crisis. The activists eventually decided to adopt TikTok and appropriated the new platform from being a "silly" one to a politically-oriented and climate-related one that was mainly used for talking with their teenage audience.

Finally, aside from leveraging social media as an instrument, activists can also wield social media platforms as weapons to attack their ideological enemies. This weaponization of social media and the internet is most evident in hacktivism, a form of activism that involves using digital tactics to attack enemies. Examples include WikiLeaks, which releases classified documents to the public, and Anonymous, a group that can be traced to the forums of 4chan and was responsible for the digital attack on the Church of Scientology in 2008. Activists can also use social media to perform bottom-up surveillance, such as filming and photographing police brutality during protests or ordinary times as a passive-aggressive tactic to monitor and expose police or state-sponsored violence. Viral videos documenting the deaths of George Floyd and Michael Brown sparked nationwide protests and contributed to the #BlackLivesMatter movement. And, in many ways, this also reflects the value of social media for archive purposes, whereby collective memories regarding specific movements get preserved via a repository of text and audio-visual content that bears permeant social and cultural values.

> **Auto-dialing Russian government officials**
>
> In the early days of the Russia-Ukraine war in 2022, a group of international hacktivists launched a web service through *WasteRussianTime.today*. The site combined prank calls and robocalls to launch telephonic annoyance on Russian officials who work in government, military, and intelligence. Interestingly, it does so by connecting two Russian officials' numbers, ostensibly wasting their time as the pranked officials try to figure out why they are called and who did it. All the while, users can listen in as the Russian officials speak. The actual effect of these calls is hard to measure; therefore, the site acted mostly as a kind of performance art installation, allowing visitors to observe and enjoy its spam calls silently.

The discussion thus far has been centered primarily on individual activists and their use of social media for collective actions. Be that as it may, activists are not the one and only stakeholder in activism. To facilitate change, particularly in the legal and political sphere, interest organizations must be involved through lobbying activities. In this process, social media might be helpful for organizational entities to shape policy debates and build public images (Chalmers & Shotton, 2016). In one study, researchers looked into the digital advocacy strategies of IF Metall, Sweden's largest trade union in the manufacturing sector with over 310 thousand members (Johansson & Scaramuzzino, 2019). To do so, researchers saved and analyzed documentation from websites and social media, including texts, photos, video clips, animations, and audio files, coupled with semi-structured interviews with people in charge of the organization's social media and media communication. The study found that IF metal uses various social media platforms, such as Facebook, Twitter, Instagram, LinkedIn, and YouTube, and customized information in those channels to reach particular types of actors/audiences, including key decision makers. For example, the union promoted on Facebook about Swedish Prime Minister's meeting with the then newly elected union leader, amplifying their use of digital access politics. The union's ties with established political parties and women's rights organizations also led them to run several digital campaigns to mobilize people for actions, such as starting a movement for equal pay. Interestingly, the researchers also observed that most digital protest activities of the union presented little direct conflict with the current government, which presumably reflected the union's tactic to win over the government's support on their cause.

Issues regarding social-media-empowered movements

Social media's potentials for protests are not without limit. In fact, some scholars and activists cautioned not to be over-optimistic toward social media; for it risks being a mere form of "clicktivism" or "slacktivism" (an epithet that combines "click"/"slacker" and "activism"), where actions are associated with relatively little time or commitment, rendering little political impact and only serving the egotistical needs of participating and feeling good about oneself (Christensen, 2011). In a similar vein, Lindgren (2013) noted that "in order for disruptive spaces to actually make a difference, and not just as sources of inspiration, identity, and mutual support in electronic isolation, they must be hybrid" (p. 150), stressing the necessity of converting online activism to offline settings. Taking a step further, popular science writer Malcolm Gladwell (2010) argued that effective social change requires hierarchical, "military-like" social structures, highlighting the unsuitability of weak ties for activism.

A second issue about using social media for activism has relevance to the *cycle effect*, in which protests often go through cycles, starting with an initial spark that leads to the rapid diffusion of participation and organization of actions, eventually leading to the downscale or dissipation of the protests. As an illustration of the cycle effect, Hensby (2016) interviewed students involved in the 2010/11 anti-austerity protests in the UK—a movement born out of the British government's agenda of reducing public sectors spending, including universities' teaching budget, warfare, and health, among others. As expected, many participants commented on how rapid and powerful social media was in contributing to raising people's awareness of the movement and grasping the attention of mainstream media. However, as the prime minister voted down the bill increasing students' fees, the movement quickly experienced a loss of attention. In this process, key participants became increasingly disconnected from the wider student body due to their use of "secret" Facebook groups to counter surveillance threats, which contradicted with the democratic spirit the movement initially set for itself.

The cycle effect also suggests that a range of external factors not directly tied to the movement may influence the direction of a given protest. Remember the Arab Spring mentioned about at the beginning of this chapter? While social media was initially credited for its role in the Arab Spring, many researchers now question its actual impact. In particular, Tufekci (2017) documented that, during the 2011 protests in Cairo's Tahrir Square, the regime shut down Internet and cell phone communication in the area, limiting protesters' communication and coordination on the ground. However, despite this obstacle, the movement ultimately led to Mubarak's resignation, partly due to prolonged media coverage worldwide and the regime's actions, such as police brutality and censorship, that escalated the confrontation. Nevertheless, critics argue that the movement failed to bring any meaningful changes to Egyptian politics, and the subsequent government was unable to effectively improve the lives of average citizens.

Some scholars also see corporate-owned social media as a factor of concern (Highfield, 2017), citing issues such as social media companies not promoting protest hashtags as trending topics, censoring certain speech or content, and leaking private communication/data to third parties or law enforcement. However, this does not mean that social media companies are inherently evil and always on the opposite side of grassroots movements. In some cases, companies engage in censorship and actions to demote protest-related hashtags and content for their own survival, inadvertently becoming part of the suppressive apparatus (Tufekci, 2017). Conversely, companies that resist government requests to censor or silence certain content may be forced to leave a particular market or simply shut down their services. For example, Google and LinkedIn both left the Chinese market due to challenging regulatory requirements, and Fanfou, one of China's earliest Twitter-like services, was shut down for an extended period due to discussions about an ethnic riot in the capital city of the Xinjiang Uyghur Autonomous Region in 2009.

Lastly, on a cultural or ideological dimension, the rise of cancel culture, or "callout culture," in recent years also reflects the growing concerns over digital activism. The term "cancel" has its 1980s pop culture origin, referring to breaking up with somebody. Today, cancel culture generally refers to a form of public backlash—often on social media—toward a public figure whose discourse and/or actions are deemed inappropriate. The canceling aspect often involves members of the online movement demanding to end the person's career or revoke one's cultural cachet (Romano, 2019). Over the years, a long list of individuals was involved, including actors, academics, athletes, influencers, corporate CEOs, comedians, politicians, as well as ordinary people. And issues that led these individuals to get canceled tend to have relevance with racism, gender equality, treatment of LGBTQ+ groups, bullying, among others.

The practice of cancel culture is often characterized as a left-leaning cultural phenomenon. To many who participate in it, calling out others on social media for their inappropriate actions is a way of holding them accountable (Vogels et al., 2021). However, some who disagree claim that cancel culture may be indicative of mob mentality and another form of censorship and harassment (Romano, 2021). Despite this, social media is often credited as the birthplace of cancel culture, as it has the ability to quickly raise awareness on issues and contribute to justice in certain cases, such as the allegations of sexual assault against Harvey Weinstein.

The design and features of social media can introduce several complications for callout movements. For instance, Bouvier (2020) conducted a critical discourse analysis on a series of race-related canceling movements on Twitter and found that the fleeting nature of these conversations and the limited in-depth discussion resulted in an "affective flow of outrage," rather than meaningful engagement with underlying structural issues. The self-presentation aspect of social media also suggests that some participants may take pleasure in "moral posturing"—a collective display of their stance on a racial issue or conviction of the alleged perpetrator—rather than addressing the deeper structural issue that may have contributed to the specific case. As a result, they may risk misrepresenting and diverting attention from the actual issue of social justice.

Social media for political campaign and election

The use of social media in political campaigns and elections often overlaps with its use in social movements, as both seek to influence and mobilize citizens through social media. Social media users are provided with opportunities to engage, organize, coordinate, and improvise using the functionalities and network structures of social media. But political campaigns and elections are also unique in that they are more top-down and less organic than social movements. Moreover, stakeholders such as political parties, politicians, strategists, and corporate donors often have more power in political campaigns than in grassroots social movements. Considering this, it is useful to pay specific attention to the role of social media in shaping the landscape of political campaigns and elections.

Before the rise of social media and other digital technologies, political campaigns were primarily conducted through traditional media outlets such as television and newspapers. However, the 2004 US presidential campaign of Howard Dean marked the debut of an Internet-based campaign. Despite ultimately withdrawing from the Democratic primary, Dean was able to raise nearly $15 million in small donations during the third quarter of his campaign, breaking Bill Clinton's 1995 record as a Democratic candidate in a presidential race. This success was largely attributed to Dean's pioneering use of the Internet, particularly blogs, for fundraising and raising awareness. In the years since, social media has become increasingly integrated into standard campaign strategies, culminating in the 2016 campaign of then-Republican presidential nominee Donald Trump, who used Twitter as his primary platform for connecting with supporters.

What emerging media technologies brought to political campaigns and elections are not just opportunities, such as the ease of accessing a critical mass, the low cost of communicating with the public (at least for minor candidates), and the efficiency of targeting and accessing specific voter blocs, but also an array of unforeseen challenges. For one, politicians today are under constant scrutiny due to social media, and any mistakes or gaffes from their past may come back to haunt them. In the 2019 Canadian election, incumbent Prime Minister Justin Trudeau faced backlash over old pictures showing him wearing blackface makeup, almost costing him his re-election. Along the same lines, rumors and attacks can spread quickly and widely on social media, as seen with the "birther movement" surrounding Barack Obama and the "Pizzagate" conspiracy about Hillary Clinton, both of which persisted long after their campaigns ended.

The second challenge that social media introduced to campaigns and elections has to do with the *delocalization* of local elections (Carr, 2020). As social media connect politics to the public, local elections involving senators and alder people become part of the larger public debate. This delocalization occurs in three main ways: financially, interactionally, and geopolitically. On the financial aspect, more and more campaigns saw financial contributions come from outside the local district. In one extreme case, Maura Sullivan, a candidate in New Hampshire's 1st Congressional District race in 2018, received 96.7% of contributions from out-of-state, non–New Hampshire donors. On the interaction aspect, social media and Internet culture encourage a unique sense of playfulness that gets incorporated into the ways that candidates interact with supporters online. In a less nefarious example, during the 2020 presidential election, one of Democratic primary candidates Andrew Yang ran a predominantly online-based campaign and was often pictured dancing and singing with his supporters on various social media platforms, which helped him develop a devoted online fan base known as the "Yang Gang" (Matthews, 2019). Finally, on the geopolitical dimension, some local political events can be so consequential that outside political forces may try to exert influence through the manipulation of social media streams and the information ecosystem. High-profile examples include, for instance, Russia's disinformation campaign aimed to sway the 2016 referendum in the United Kingdom for Britain to exit the European Union and the secret-spilling website WikiLeaks' release of Hillary Clinton's email archive, which had an outsized impact on her campaign outcome.

Last but not least, social media also present challenges to the way candidates communicate with their constituents. In the era of digital campaigns, social media provided a new channel for political candidates to construct and negotiate their own public persona. In this process of personalization, the role of authenticity becomes one of the most essential elements for voters to gauge politicians' character and trustworthiness. Perhaps no other case can make the point as resounding as the 2016 US presidential election cycle. Through an analysis of Clinton and Trump's tweeting style, Enli (2017) found that nearly 82% of Clinton's tweets were considered professionally produced and adhering to the conventional standards as a presidential candidate, whereas more than half (55%) of Trump's tweets were amateurish, with haphazard taunts such as "delete your account." Clearly, Trump's spontaneity made it much easier to claim authenticity in front of his supporters. This point was echoed by Clinton in her post-election memoir *What Happened* (2017), where she wrote:

> In any case, this whole topic of 'being real' can feel very silly. . . . Yet the issues of authenticity and likability had an impact on the most consequential election of our lifetime, and it will have an impact on future ones.
>
> (p. 124)

To be sure, candidates' images have always been a central focus of political campaigns, but social media has complicated this aspect of campaigning, particularly for younger voters who value honesty and authenticity. Candidates who can effectively present a socially attractive persona on social media platforms are more likely to attract attention (Enli & Rosenberg, 2018). This is partially supported by experimental research, such as a study that recruited online participants to view Instagram images posted by two 2020 US Congressional candidates (one male and one female) and record their reactions (Zulli & Towner, 2021). These images varied in terms of "liveness" (showing up at live activities), "authenticity" (revealing private life and personality), and "emotionality" (projecting positive or negative feelings). The results showed that participants who viewed authentic images of both candidates (one showing the candidate with her mother, the other showing the candidate with his son) were more likely to vote for them than those who viewed neutral images of the candidates blended in with a crowd.

Changing campaign strategies

Aside from its impact on politicians and candidates, the integration of social media into modern political campaigns has also led campaign strategists and publicists to reassess their media strategies. These individuals play a crucial role in most election campaigns, with some, such as Joe Rospars and Steve Bannon, becoming well-known for their tactics. Rospars worked on Howard Dean's 2004 campaign and later became a key member of Barack Obama's successful 2008 and 2012 campaigns. Bannon, co-founder of the far-right news website *Breitbart News*, served as Chief Strategist for Donald Trump's 2016 campaign and was known for his "flooding the zone" strategy, which aimed to confuse and distract the mainstream media and their audience by constantly releasing potentially relevant or irrelevant information. As such, how these strategists see social media must be closely examined.

In an insightful research project, Kreiss and colleagues (2018) conducted a series of in-depth qualitative interviews with campaign professionals active during the 2016 presidential cycle. The interviewees include the social media manager, deputy campaign manager, and digital strategist for major candidates such as Hillary Clinton, Jeb Bush, Ted Cruz, and Bernie Sanders. Analysis of the interviews revealed that campaign professionals are acutely aware of the subtle differences across social media platforms and emphasized the necessity of tailoring their digital content based on audience demographic characteristics and cultural norm in each specific platform (although this is not always the case in practice). Here, understanding who the audiences are is essential for these professionals, and they rely on the analytics services offered by social media platforms and other proprietary software for such information. Interestingly, campaign staff members also acknowledged that the affordances of the social media platforms heavily affect their decisions about how to approach each platform. For example, a manager of Jeb Bush's campaign remarked that the difficulty of using outside links and sharing content among users largely limited Instagram's and Snapchat's electoral value. Others suggested that the anonymity of Snapchat also made it hard for campaigns to convert users to voters or donors.

However, when necessary, campaign managers can find ways to work around a specific platform's design and affordance restrictions. For instance, when dealing with platforms that use algorithms to filter users' news feeds, campaign managers can override these algorithms by using pay-to-promote services such as "boosting" on Facebook. In contrast, social media platforms that are less influenced by algorithmic filtering, such as Snapchat, WhatsApp, and Telegram, are much easier and cheaper to use for message dissemination (Bossetta, 2018). Platform policies can also pose design limits, such as Twitter's ban on all paid political advertising in 2019, making it the first major social network site to do so and creating a stark contrast with Facebook (Conger, 2019). Nevertheless, savvy campaign managers can still bypass these platform policy constraints by organizing candidates' supporters on one platform to promote content on another. As a result, a qualified social media campaign manager must continually update their knowledge about each platform's features, functionalities, and rules to develop creative strategies that can make a real difference.

Changing campaign engagement

In the broadcast era, political campaigns tend to put the audience on the receiving end and thus consider audience members mere receivers of campaign messages. Social media changed that convention by letting everyday people join the campaign and have their say. No longer are audiences passively accepting candidates' campaign messages; rather, they co-create with politicians and campaign strategists through performance, response, rebuke, and meaning-making across multiple platforms.

A classic example took place in January 2021: as citizens of the entire nation were watching the inauguration of President Joe Biden, a photo of Vermont senator Bernie Sanders featured him masked and cross-legged, sitting on a folding chair with a bulky coat and mittens in the audience seats suddenly went viral on the Internet. Supporters and non-supporters alike went for a creativity streak, appropriating memes showing the senator as a member of the *Avengers* or on the Iron Throne from *Game of Thrones*. This, coupled with Sanders' popularity among young voters, further strengthened his image as a down-to-earth populist politician who drastically differs from many of his peers. Other news and entertainment media joined the bandwagon by digging the backstories about Sanders' mitten, creating even more favorable online discussions around the former Democratic primary candidate. Again, instances like this are numerous and indicate that social media users are playing active roles in political campaigns, which can be vital for candidates to obtain media exposure and harness support.

If appropriating memes online is one way for voters to engage asynchronously with political campaigns and elections, "second screening" or "live-tweeting" represents a novel approach for voters to participate in political campaigns in real-time. In the United States, debates during the primary and national stages of elections are popular events that traditionally generate high television ratings. However, with the advent of social media and mobile devices, many users now watch these debates on television while simultaneously interacting on various social media platforms. This engagement includes activities such as sharing reactions, fact-checking candidates' statements, connecting with like-minded individuals, and commenting on candidates or other users. Such behavior combines real-time audience engagement with on-stage events and opens the door to political participation, audience analytics, and journalistic reporting. Studies conducted in

Figure 10.2 Supporters hold a banner memeified to represent Andrew Yang's "Freedom Dividend" proposal

(*Source*: Janson George/Shutterstock)

various democratic countries have shown that this behavior significantly enhances users' enjoyment of debates, level of engagement, intention to vote, and knowledge of candidates' policies. Ultimately, this contributes to democratic citizenship and political equality (Gil de Zúñiga & Liu, 2017; Vaccari & Valeriani, 2018).

Realizing the power and value of social media for political campaigns, platforms such as Twitter provides an official guide regarding how to use its services for precisely campaign relative events. On their website, users will find "Workbook for live Tweeting" as well as "Campaigning on Twitter: The Handbook for NGOs, Politics & Public Service," instructing campaign strategists and social media managers to engage users during live political events. Together, the symbiosis between platforms, social media users, candidates, and campaign strategies made modern campaign elections much more sophisticated and intricate.

Key issues in social-media-driven campaigns and elections

The increasingly social-media-mediated campaign elections are not without democratic consequences and social concerns. Some of the considerations overlap with the changes social media brought to the landscape of news and information sharing, which we have covered extensively in Chapter 9. Here, the discussion will mainly focus on social media's impact in the political campaign realm.

First, recent US presidential elections, including the highly controversial 2016 election, which saw significant interference from Russian sources, have brought the concept of bots to the forefront of public discourse. Traditionally known among tech workers, bots refer to software robots, created to assist in completing specific tasks for humans. A well-known example of bots is chatbots—computer algorithms programmed to interact with humans (such as those encountered when reaching out to an internet service provider, or simply, ChatGPT).

Despite their seemingly benign origins, a new type of bot—social bots—has emerged in various social media platforms over the past decade, with the aim of emulating and potentially influencing people's behavior through automatically produced content and interactions. Social bots are (semi-)automated accounts on social media that can perform tasks such as liking or sending friend requests, posting or sharing predefined content, crawling the social media platform for predefined keywords, and extracting user data (Assenmacher et al., 2020). Many argue that the massive use and deployment of social bots to spread political opinions or misinformation campaigns pose an imminent threat to democratic elections in the US and European countries (Ferrara, 2017). For example, during the 2016 presidential election, Twitter uncovered more than 50,000 Russia-supported bots, which reached more than 1.4 million Americans. Needless to say, manually banning or blocking these bots will be futile. Therefore, how to efficiently detect social bots is an emerging research field, with researchers actively testing various techniques to identify bots on each platform (Martini et al., 2021). The good news is that majority of the social bots on social media show a limited degree of intelligence at present (Assenmacher et al., 2020), but with the wide adoption of the machine and deep learning technologies, it may be a matter of time before these social bots become "intelligent." Or, perhaps they are already among us.

The second issue that social media brought to the political sphere is the increasingly intensified polarization within both political elites and voters (Klein, 2020). Here, social media is often considered a magnifier for the matter because its affordance favors viral and sentimental discourses more than calm and sophisticated conversations. To test this hypothesis, Brady and colleagues (2021) built a machine-learning software capable of tracking moral outrage in Twitter posts and used it to perform a couple of observational studies over 12.7 million tweets from 7,331 Twitter users. They found that users who received more "likes" and "retweets" when they expressed outrage in a tweet were more likely to express outrage in later posts, suggesting that the design incentives such as Twitter really do change how people discuss issues online. What is more, they also performed a set of experiments where participants were randomly assigned to scroll a simulated Twitter feed with either a majority of outraged expressions (and these outraged expressions displayed more likes and shares than the neutral tweets) or a majority of neutral expressions. After the scrolling session, these participants were then set to complete a number of trials where they are asked to choose between an outrage and a neutral tweet to post on the simulated Twitter network. The results suggested that members of politically extreme networks expressed more outrage than members of politically moderate networks. However, members of politically moderate networks were actually more influenced by social rewards. In the long run, and at a massive scale, this intensified outrage may lead ordinary Americans to dislike and distrust those from the other party, leading to a phenomenon of hyper-animosity between the parties known as affective polarization (Iyengar et al., 2019). Nonetheless, it would be a mistake to think political polarization on social media is a unified phenomenon because of significant cross-platform differences. While Twitter may encourage inter-group hostility, other platforms such as WhatsApp have shown signs of depolarization over time (Kligler-Vilenchik et al., 2020). Once again, this suggests that how we design platforms matters for the health of civil society.

Reinvigorating Populism?

The global political landscape has witnessed a surge in populist movements in recent years, with prominent figures like Donald Trump and Bernie Sanders in the US, Marine Le Pen in France, Giorgia Meloni in Italy, Jair Bolsonaro in Brazil, and Viktor Mihály Orbán in Hungary gaining significant support in their respective domestic elections. Some experts posit that social media's ubiquity in political campaigns, professional journalism, and political debates has fueled this trend, with politicians who employ rhetoric that pits "us, the people" against "them, the elites" gaining an upper hand. For instance, a case study of the 2016 US presidential debate showed that Donald Trump's confrontational communication style, including personal attacks, anger, and blame, attracted more attention on Twitter than his opponent Hillary Clinton (Bucy et al., 2020). However, a recent study that analyzed data from 27 countries during the COVID-19 pandemic and 79 sources covering 8 million individuals since 1958 revealed that the pandemic seemed to reverse the rise of populism. In many countries, support for populist parties, approval of populist leaders, and agreement with populist attitudes decreased (Foa et al., 2022). Nonetheless, it remains to be seen if this reversal will continue.

Lastly, and as we have touched on before, the reliance on social media also ushered in the era of big-data-fueled campaigns, where campaign strategists and publicists increasingly rely on big data and social media analytics to monitor public opinions and to adjust advertising campaigns at critical moments. At the industry level, this means that social media companies are in an arms race to offer the most compressive user data (including demographics and user actions on their platforms) to campaign strategists. Once again, as users, we are in our familiar territory of privacy invasion, where our rights to own our own online data, corporate interest, and user consent comingle. In a heavily publicized case, Cambridge Analytica—a British political consulting firm—collected millions of Facebook users' data for generating tailored political advertisements during the 2016 presidential campaigns of Ted Cruz and Donald Trump. The scandal prompted a massive public outcry for more stringent data regulation for political campaigns, sending shockwaves to the social media and campaign analytics industry that rely on users' data for financial gain. In the wake of the scandal, Facebook agreed to pay a hefty fine of nearly $5 billion to the Federal Trade Commission due to its privacy violation, making it one of the most significant penalties ever assessed by the US government for any violation (Nylen, 2021). However, social media legislations about political advertising and data protection remain stalled in the US. New regulating policies are highly contested in the court and Congress, and analytics continue their business as usual, more or less. As such, the impact of social media looms ever larger in future elections.

Keywords: Connective action; clicktivism; slacktivism; the cycle effect; delocalization; personalization; second-screening; cancel culture; bots; political polarization; affective polarization; populism.

References

Assenmacher, D., Clever, L., Frischlich, L., Quandt, T., Trautmann, H., & Grimme, C. (2020). Demystifying social bots: On the intelligence of automated social media actors. *Social Media + Society*, 6(3). https://doi.org/10.1177/2056305120939264

Belotti, F., Donato, S., Bussoletti, A., & Comunello, F. (2022). Youth activism for climate on and beyond social media: Insights from FridaysForFuture-Rome. *The International Journal of Press/Politics*, 27(3), 718–737.

Bennett, W. L., & Segerberg, A. (2013). *The logic of connective action: Digital media and the personalization of contentious politics*. Cambridge University Press.

Bossetta, M. (2018). The digital architectures of social media: Comparing political campaigning on Facebook, Twitter, Instagram, and Snapchat in the 2016 US election. *Journalism & Mass Communication Quarterly*, 95(2), 471–496.

Bouvier, G. (2020). Racist call-outs and cancel culture on Twitter: The limitations of the platform's ability to define issues of social justice. *Discourse, Context & Media*, 38, 100431. https://doi.org/10.1016/j.dcm.2020.100431

Brady, W. J., McLoughlin, K., Doan, T. N., & Crockett, M. J. (2021). How social learning amplifies moral outrage expression in online social networks. *Science Advances*, *7*(33), eabe5641. https://doi.org/10.1126/sciadv.abe5641

Bucy, E. P., Foley, J. M., Lukito, J., Doroshenko, L., Shah, D. V., Pevehouse, J. C. W., & Wells, C. (2020). Performing populism: Trump's transgressive debate style and the dynamics of Twitter response. *New Media & Society*, *22*(4), 634–658. https://doi.org/10.1177/1461444819893984

Cammaerts, B. (2015). Social media and activism. In R. Mansell & P. Hwa (Eds.), *The international encyclopedia of digital communication and society* (pp. 1027–1034). Wiley-Blackwell. https://doi.org/10.10hal02/9781118767771.wbiedcs083

Carr, C. T. (2020). The delocalization of the local election. *Social Media + Society*, *6*(2), 2056305120924772.

Chalmers, A. W., & Shotton, P. A. (2016). Changing the face of advocacy? Explaining interest organizations' use of social media strategies. *Political Communication*, *33*(3), 374–391.

Christensen, H. S. (2011). Political activities on the internet: Slacktivism or political participation by other means? *First Monday*, *16*(2). https://doi.org/10.5210/fm.v16i2.3336

Clinton, H. R. (2017). *What happened*. Simon & Schuster.

Conger, K. (2019, October 30). Twitter will ban all political ads, C. E. O. Jack Dorsey says. *The New York Times*. www.nytimes.com/2019/10/30/technology/twitter-political-ads-ban.html

Enli, G. (2017). Twitter as arena for the authentic outsider: Exploring the social media campaigns of Trump and Clinton in the 2016 US presidential election. *European Journal of Communication*, *32*(1), 50–61.

Enli, G., & Rosenberg, L. T. (2018). Trust in the age of social media: Populist politicians seem more authentic. *Social Media + Society*, *4*(1), 2056305118764430.

Ferrara, E. (2017). Disinformation and social bot operations in the run up to the 2017 French presidential election. *First Monday*, *22*(8). https://doi.org/10.5210/fm.v22i8.8005

Foa, R. S., Romero-Vidal, X., Klassen, A. J., Fuenzalida Concha, J., Quednau, M., & Fenner, S. (2022). The great reset: Public opinion, populism, and the pandemic. *Centre for the Future of Democracy*. https://doi.org/10.1353/jod.2022.0010

Gil de Zúñiga, H., & Liu, J. H. (2017). Second screening politics in the social media sphere: Advancing research on dual screen use in political communication with evidence from 20 countries. *Journal of Broadcasting & Electronic Media*, *61*(2), 193–219. https://doi.org/10.1080/08838151.2017.1309420

Gladwell, M. (2010, October 4). Small change: Why the revolution will not be tweeted. *The New Yorker*. www.newyorker.com/magazine/2010/10/04/small-change-malcolm-gladwell

Gleason, B. (2013). # Occupy wall street: Exploring informal learning about a social movement on Twitter. *American Behavioral Scientist*, *57*(7), 966–982. https://doi.org/10.1177/0002764213479372

Hensby, A. (2016). Campaigning for a movement: Collective identity and student solidarity in the 2010/11 UK protests against fees and cuts. In *Student politics and protest* (pp. 31–48). Routledge.

Highfield, T. (2017). *Social media and everyday politics*. John Wiley & Sons.

Iyengar, S., Lelkes, Y., Levendusky, M., Malhotra, N., & Westwood, S. J. (2019). The origins and consequences of affective polarization in the United States. *Annual Review of Political Science*, 22(1), 129–146. https://doi.org/10.1146/annurev-polisci-051117-073034

Johansson, H., & Scaramuzzino, G. (2019). The logic of digital advocacy: Between acts of political influence and presence. *New Media & Society*, 21(7), 1528–1545. https://doi.org/10.1177/1461444818822488

Klein, E. (2020). *Why we're polarized*. Avid Reader Press.

Kligler-Vilenchik, N., Baden, C., & Yarchi, M. (2020). Interpretative polarization across platforms: How political disagreement develops over time on Facebook, Twitter, and WhatsApp. *Social Media + Society*, 6(3), 2056305120944393.

Kreiss, D., Lawrence, R. G., & McGregor, S. C. (2018). In their own words: Political practitioner accounts of candidates, audiences, affordances, genres, and timing in strategic social media use. *Political Communication*, 35(1), 8–31.

Li, M., Turki, N., Izaguirre, C. R., DeMahy, C., Thibodeaux, B. L., & Gage, T. (2021). Twitter as a tool for social movement: An analysis of feminist activism on social media communities. *Journal of Community Psychology*, 49(3), 854–868. https://doi.org/10.1002/jcop.22324

Lindgren, S. (2013). The potential and limitations of Twitter activism: Mapping the 2011 Libyan uprising. *tripleC: Communication, Capitalism & Critique. Open Access Journal for a Global Sustainable Information Society*, 11(1), 207–220. https://doi.org/10.31269/triplec.v11i1.475

Lyons, K. (2021, January 22). *GameStop stock halts trading after Reddit drama*. www.theverge.com/2021/1/22/22244900/game-stop-stock-halted-trading-volatility

Martini, F., Samula, P., Keller, T. R., & Klinger, U. (2021). Bot, or not? Comparing three methods for detecting social bots in five political discourses. *Big Data & Society*, 8(2). https://doi.org/10.1177/20539517211033566

Matthews, D. (2019). *Andrew Yang, the 2020 long-shot candidate running on a universal basic income, explained*. www.vox.com/2019/3/11/18256198/andrew-yang-gang-presidential-policies-universal-basic-income-joe-rogan

Nylen, L. (2021). Facebook paid billions extra to the FTC to spare Zuckerberg in data suit, shareholders allege. *Politico.com*. www.politico.com/news/2021/09/21/facebook-paid-billions-extra-to-the-ftc-to-spare-zuckerberg-in-data-suit-shareholders-allege-513456

Romano, A. (2019). Why we can't stop fighting about cancel culture. *Vox.com*. www.vox.com/culture/2019/12/30/20879720/what-is-cancel-culture-explained-history-debate

Romano, A. (2021). *The second wave of "cancel culture"*. www.vox.com/22384308/cancel-culture-free-speech-accountability-debate

Tufekci, Z. (2017). Twitter and tear gas. In *Twitter and tear gas*. Yale University Press. https://doi.org/10.1007/s10767-019-9317-2

Urman, A., Ho, J. C. T., & Katz, S. (2021). Analyzing protest mobilization on Telegram: The case of 2019 anti-extradition bill movement in Hong Kong. *PLoS One*, 16(10), e0256675. https://doi.org/10.1371/journal.pone.0256675

Vaccari, C., & Valeriani, A. (2018). Dual screening, public service broadcasting, and political participation in eight western democracies. *The international Journal of Press/Politics*, 23(3), 367–388. https://doi.org/10.1177/1940161218779170

Vogels, E., Anderson, M., Porteus, M., Baronavski, C., Atske, S., Mcclain, C., Auxier, B., Perrin, A., & Ramshankar, M. (2021). *Americans and 'cancel culture: Where some see calls for accountability, others see censorship, punishment.* www.pewresearch.org/internet/2021/05/19/americans-and-cancel-culture-where-some-see-calls-for-accountability-others-see-censorship-punishment/

Zulli, D., & Towner, T. L. (2021). The effects of "live," authentic, and emotional instagram images on congressional candidate evaluations. *Social Media + Society*, 7(4), https://doi.org/10.1177/20563051211062917

11 Social media marketing

Remember the old days when marketing and branding followed a different playbook? Back then, all you needed was a solid product that met some basic needs and a 30-second TV commercial during prime-time on a national network, and your brand would be etched in the minds of millions for years to come. After all, that's how sports fans around the globe came to recognize the Swoosh and the phrase "just do it" and how the green lizard became synonymous with saving you 15% in 15 minutes on insurance.

But, as more and more brands and companies entered the market, vying for limited media resources and the dwindling attention span of audiences, traditional marketing tactics hit a wall. Despite an influx of ads in the form of sponsored editorials, product placement, and even banana peels, audiences have become increasingly indifferent to marketing messages, often skipping or blocking ads. This phenomenon is compounded by the fragmentation of audiences, who are developing their own unique tastes and needs and splintering into smaller and smaller segments across various media offerings. And thus, a perfect storm was born, opening the door for the rise of social media and the Internet.

The advent of social media fundamentally changed the marketing world, and for good reasons. Social media offers companies and brands three main things: electronic word-of-mouth (eWOM), or peer-to-peer marketing, two-way communication, and narrowcasting (as opposed to "broadcasting"). First, interpersonal communication, such as word-of-mouth from peer consumers, has been found to be a highly effective and impactful form of advertisement. This is partly why multi-level marketing, despite its controversial reputation, remains a successful strategy (Erkan & Evans, 2016). Second, social media's two-way communication allows for communication between brands and their followers, highlighting the importance of interactivity in marketing efforts (Ariel & Avidar, 2015). Lastly, as consumers become more fragmented in their tastes and needs, the traditional approach of broadcasting unified advertisements to a mass audience is no

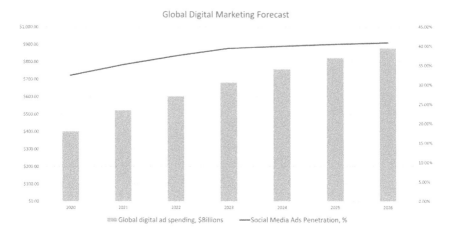

Figure 11.1 Global digital marketing forecast
(*Data Source*: nasdaq.com)

longer effective. Instead, there is a need for a sense of exclusivity, with the brand speaking directly to unique customers playing a crucial role in marketing success. By establishing a social media presence on multiple platforms, brands can reach users of different interests and demographics with targeted, cost-effective messages (Berge, 2014).

Social-media-marketing triad

Today, American companies spend on average 15% of their marketing budgets on social media. But, exactly how do modern brands and companies leverage the advantages of social media for marketing endeavors? And how effective are they? The remainder of the section will approach these questions through the lens of the marketing triad (Burcher, 2012; Colicev et al., 2018): earned media (i.e., promotion-driven publicity); owned media (i.e., content channels directly controlled by marketers); and paid media (i.e., the traditional purchase of ad time and space).

Paid media

Let us start with the most familiar one, paid media. Paid media originated from the conventional marketing practices in the traditional media space, referring to the direct purchase of advertising time and space to increase brand awareness. On social media, paid media practices are manifested by brands directly buying spaces on platforms to display ads, including banners on the side, sponsored updates in the news feed, and even targeted ads on smart TVs and other emerging media platforms (Wolk, 2019). Against this backdrop, sophisticated consumer marketing and advertising research, including the use of eye-tracking and other psychophysiological measurement devices, becomes increasingly crucial for brands and companies to learn about the most effective ways for advertisement placements.

> **Exercise**
>
> Picture this: as you walk around a department store, a QR code sits next to a nice designer bag. A quick scan of the code brings up a statue of Elphis, the Greek god, which then walks around in your surroundings as you move your camera, and you will be able to take photos or even videos with it and share them with your friends. Yes, that is an augmented reality advertisement (from Burberry UK). As mobile AR users grow worldwide, so does the AR advertisement industry. Unlike virtual reality, which completely immerses a user in a computer-generated environment, AR enables the user to view the real world upon which computer-generated objects are superimposed, thereby enhancing reality rather than replacing it.
>
> In one of the first studies about AR advertising, Scholz and Smith (2016) identified four typical augmented reality marketing paradigms. The first, called Active Print/Packaging, involves the use of a smartphone, tablet, or computer camera to scan AR-equipped printed materials (such as print ads, product packaging, or catalogs) in order to access digital content displayed on the mobile device or computer screen. The second, called Bogus Window, involves the use of devices such as TV screens disguised as standard glass windows to alter the space in view of the user with digital objects. The third, called Geo-Layer, involves the augmentation of the space around the user with digital objects that may or may not be linked to specific locations. The fourth, called Magic Mirror, involves the use of an AR mobile application or an AR-equipped screen to interact with virtual objects and see oneself as part of the augmentation.
>
> Now that you are familiar with these four types of AR ads by definition, try to find an example for each. Take mental notes on how you react to these ads. Then, share your examples with your friends and colleagues and see what works and what does not. Do you react to these ads similarly or differently? What makes a good AR advertisement, in your view? Finally, have you encountered any latest AR advertisement that don't fit into these existing typologies? If so, how would you describe this new type that you have seen and experienced?

Earned media

Compared with paid media, earned media appear to be less invasive. Simply put, earned media refers to publicity that brands and companies do not pay for directly. This type of publicity is extremely valuable, as it resembles powerful offline word-of-mouth promotion, albeit in digital form. Depending on the platform where earned media is distributed, it can take different forms, such as TikTok unboxing videos, Amazon or Google product reviews, or even Reddit shopping tips. Consumers tend to trust earned media more because it often appears authentic and organic. Moreover, many social media users voluntarily participate in brand publicities, such as factory trips to the World of Coca-Cola or the Budweiser Brewery Experience, which typically result in countless social media photos and videos being spread in virtual networks, big and small, young and old. To companies, these types of content online are essentially free advertisements, reaching millions of customers in seconds without the jaw-dropping price of a Super Bowl commercial.

Owned media

Brands and companies also rely on their own media channels, such as company-owned official or verified social media accounts and fan pages, to disseminate marketing messages. Studies have shown that direct communication with consumers via these official channels can increase consumers' knowledge of brand features and benefits as well as enhance their engagement with the brand. Depending on the product category, these marketing efforts may even contribute directly to sales revenue. For example, in one study about the efficacy of owned media, researchers collaborated with a leading travel agency in Taiwan (Chang et al., 2018). They collected sales data of their international travel products that had an advertisement on the agency's own Facebook page, matched with products that received no Facebook publicity. Results showed that the agency's owned media-marketing activities directly increased the sales of their tourism products.

Today, owned media-marketing practices are typically managed by companies' in-house marketing and communication teams, with designated social media teams handling direct communications with customers across various digital platforms. Increasingly, however, the public presence of a company's top executives on social media is drawing significant attention from the public. Their spontaneous social media discourses could directly affect the public perception of the brand and its products. A felicitous example of this is Tesla, where its CEO Elon Musk's social media presence is so palpable that it can affect the company's stock price and market performance.

In addition to direct messages from the brands and their representative, owned media practices also involve official marketing campaigns and initiatives conducted via social media. By doing so, brands invite consumers to be co-creators, ostensibly empowering people to be part of a larger "movement." There are two main types of these practices: first, brands encourage customers to take selfies or share product experiences using designated company hashtags (e.g., Coca Cola's #shareacoke, Naked juice's #DrinkGoodDoGood, Benefit's #Realsies); second, brands initiate socially-oriented campaigns (or, Corporate Social Responsibility campaigns) for environmental causes (Patagonia's Buy Less, Demand More), social justice (Nike's Dream Crazy), or health causes (Yoplait's Save Lids to Save Lives).

While these campaigns are launched through the brands' own channels (i.e., official accounts), they contribute to marketing efforts in different ways. The former type emphasizes inviting users to share their own encounters with the products through social media messages; hence, the name user-generated content (UGC) marketing. To examine the actual effect of UGC marketing, for example, Müller and Christand (2019) recruited 156 participants interested in video games to watch a gaming video shown on a YouTube Channel either disclosed as sponsored advertising content or as an independent review. Results suggested that, compared to the sponsored content condition, participants who viewed the game review in the UGC condition had a more positive attitude toward the game. This is because UGC lowers potential consumers' recognition of the content as an advertisement, subsequently increasing their trust toward the content, resulting in a higher brand attitude. In other words, UGC marketing works partly because consumers do not see it as a blunt advertisement.

It is important to note that companies do not always have complete control over user-generated content. If a brand's UGC marketing efforts result in negative reviews or comments on social media, the risks can outweigh any potential benefits. One prime example is the "#McDStories" campaign launched by McDonald's in 2012. The campaign encouraged customers to share their positive experiences at McDonald's restaurants using the hashtag #McDStories. However, many of the responses were negative, with customers sharing stories of poor service, dirty restaurants, and subpar food quality. This led to a significant backlash against the brand on social media, damaging its reputation and causing the campaign to be quickly shut down.

In comparison to UGC marketing, Corporate Social Responsibility (CSR) marketing aims at cultivating favorable awareness by establishing a framework of stakeholder relationships, particularly between consumers and brands. Such efforts are theoretically viable because more and more consumers nowadays are aware of and invested in various social causes, and thus, expect companies and brands to play active roles in addressing social and environmental issues through minimizing environmental impacts, voluntarism, corporate philanthropy, and taking a stance on social problems such as equality, diversity, human rights, and education. Companies that are able to align their own social values with consumers' interests and communicate that effectively with consumers would be favored more than those who fail to do so. Indeed, research suggested that Fortune 500 companies who are more apt at communicating their CSR efforts on social media appear to be more effective in attracting followers and generating social media discussions (in replies, mentions, and retweets) than companies who aren't as good at CSR communication (Saxton et al., 2019).

Of course, CSR marketing is easier said than done, for the success of companies' CSR marketing endeavors is often a result of a multitude of factors (Fernández et al., 2022), such as CSR fit (fitness of a company's CSR initiatives with the company activities and customers), CSR empowerment (companies can provide consumers with control over their activities), humane-oriented appeals (emotional communication appeals of the CSR message), social media endorsement and opposition (CSR message liking, commenting, and sharing), and the sharing source (non-corporate vs corporate CSR information source). From a cross-cultural point of view, CSR marketing is also shaped by its overarching cultural values, political and economic system, and development stage. Thus, it is no surprise that corporations' CSR practice and communication, as well as consumers' perceptions and responses toward CSR, exhibit varying features in different societies. For example, one survey research shows that, in terms of consumers' engagement with social media CSR communication, Chinese consumers are shown to be influenced by their level of being an opinion leader and the extent to which they rely on communication with peers for decision-making, whereas American consumers' engagement with CSR were agnostic to these factors (Chu et al., 2020).

Marketers also need to understand meaningful CSR communication should be firmly grounded in the company's business plan and backed by corporate and employee conduct, not just an act of symbolic communication. A company that chants human rights protection on Twitter but employs child labor in its production is most likely to be seen as ingenuine and even hypocritical. A fashion company that advocates for environmental protection and ethical consumption but generates huge waste and toxic emissions will be accused of greenwashing and eventually backfire on their campaign efforts (Einwiller et al., 2019).

> **Exercise**
>
> Companies' CSR messages on social media exhibit diversity in their focus and creativity in their presentation format. As a result, the sheer variety of CSR messages can be overwhelming. For instance, companies post messages related to environmental sustainability, charitable organizations, communities, social causes, and responsible corporate practices such as fair hiring practices. To systematically categorize and analyze CSR messages, Nave and Ferreira (2019) reviewed 25 years' worth of CSR research literature and developed four criteria: (a) dimensions, (b) benefits, (c) value creation and stakeholders, and (d) motivations.
>
> The dimensions criterion emphasizes the general topics that companies choose to focus on, which defines their ethics and sustainability. Benefits concern potential gains from companies' CSR messages, including reputation, performance, and competitive advantage. Stakeholders are categorized into internal and external, referring to employees and managers of the company, and customers and society at large, respectively. Motivations include strategic and altruistic ones, concerning companies' managerial purposes and the moral value of benefiting other parties.
>
> With this set of CSR message analytic criteria, take a quick survey of your social media feed and identify any CSR messages. Take screenshots for several of these messages and apply the criteria you just learned to examine them. Can you differentiate among them easily? What can you learn from this process?

Social media influencers

In the era of social media marketing, one practice that has seen rapid growth is influencer marketing, which involves working with popular social media personalities to promote a brand's products or services to their followers. According to Influencer Marketing Hub, a Danish marketing-research company, the influencer-marketing industry was worth $16.4 billion in 2022, with 93% of marketers using influencer marketing (Santora, 2022). But do influencers really work as a specific form of social media marketing? Theoretically, the answer is yes. According to the two-step flow of communication model (Katz & Lazarsfeld, 1955), there are opinion leaders in every social network. Since interpersonal relationships and word-of-mouth communication are generally more trusted than mass media, opinion leaders serve as key intermediaries in the flow of information between message senders and receivers. Alternatively, the concept of *parasocial relationships* can also explain the efficacy of social media influencers. In media and communication literature, parasocial relationships refer to one-sided relationships that audiences develop with media figures (Horton & Wohl, 1956). And, in the context of social media use, repeated exposure to social media influencers may lead consumers to a sense of deep bond with the influencers.

Interestingly, audience members don't just develop parasocial relationships with *real* media figures. Virtual influencers, such as digital avatars or characters, present new opportunities for brands to reach and engage with consumers. Recent evidence suggested that consumers would interact and respond to virtual influencers just as they do with their human counterparts (Brown, 2020). As the digital technologies behind virtual influencers becoming more and more powerful and readily accessible, there stand a good chance for virtual influencers to be a reliable marketing tool for brands.

Factors making social media influencers click

As consumers, we don't trust social media influencers blindly. In some situations, as influencers grow their impact and followers on social media, they may become distant and less trusted by some of their followers. Therefore, it is pivotal to understand how influencers can foster parasocial relationships to generate commercial impact. On that, a sizeable body of research suggests that the key is to establish and strengthen the influencer's authority, credibility, and reliability (Lou & Yuan, 2019).

Consumers are more likely to trust influencers' product recommendations if the influencers have consistently proven themselves to be experienced and knowledgeable in certain topics and subject areas. To some extent, this is why the fit between the influencer and the endorsed product matters (Schouten et al., 2020): a beauty influencer who endorses tech products, or a fitness influencer who recommends food and beverages, may be viewed as untrustworthy, while a fashion influencer conducting clothing hauls may be deemed reasonable and trustworthy. However, influencers' credibility is also influenced by many other peripheral factors. For instance, many marketing scholars have found attractiveness to be an important ingredient in building influencer credibility (Chekima et al., 2020; Kim & Kim, 2021). Here, physical appearance, such as facial beauty, height, weight, and even skin color, is often emphasized, which may explain the prevalence of young, White, cosmopolitan, and physically fit influencers.

Beyond the obvious need for credibility, social media influencers also have to be authentic, relatable, and accessible; for that is how they win over followers in the first place. On the surface level, consumers are almost always looking for influencers who are similar to themselves (in terms of gender, age, ethnicity, professional background, or even personality). This also explains why traditional celebrities that appear to be lofty and distant are rather difficult for consumers to be identified with (Schouten et al., 2020). Savvy influencers know how to engineer authenticity and relatability. In a study about #fitpo (a.k.a. fitness inspiration; also see Chapter 8) influencers, for instance, Reade (2021) found that fitness influencers generate authenticity and relatability in three significant ways; namely, (1) posting the body raw, (2) storying the everyday, and (3) "real talk" about topics such as body image and mental health. In particular, the researcher found that, as the social media culture evolves, fitness influencers began to see the unedited and imperfect visual display of the body as a form of aesthetic, and that emotional labor ostensibly works to cultivate digital intimacies between the influencers and their Instagram followers.

When it comes to engaging with brands and sponsored content, influencers sometimes worry that being too blunt in disclosing brand sponsorship might push their followers away. The perceived commercial orientation by their followers would erode fans' trust towards the influencers (Martínez-López et al., 2020). More specifically, sponsorship disclosure messages such as the hashtag "sponsored" can trigger consumers to speculate that the influencer may receive compensatory rewards and has ulterior motives for posting the product. This eventually leads to advertising recognition and the formulation of negative product attitudes (Kim & Kim, 2021). Does that imply that influencers should approach sponsored content subtly? Promoting brands and products in a humble manner, as in humblebragging—indirectly boasting oneself (or the sponsorship) through covert ways such as complaints—is not the answer. Studies have shown that influencers who use humblebragging in the context of social media brand endorsement decrease consumers' positive brand attitude. In contrast, celebrities' humblebragging led to a reverse-positive effect on consumers' brand attitude (Paramita & Septianto, 2021).

Despite the common belief that influencers are there to influence, and the larger their following, the better, a trend in the influencer marketing world is that more and more brands are showing interest in micro-influencers—content creators with 1,000 to 10,000 followers (Santora, 2022), rather than those with over a million followers (i.e., mega-influencers). This practice is supported by empirical evidence. Through a series of experiments, researchers have consistently demonstrated that endorsements from micro-influencers are more effective than those from mega-influencers in terms of

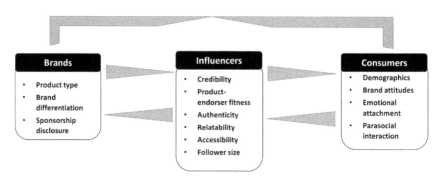

Figure 11.2 An integrative framework of social media influencers

increasing consumers' purchase intention (Park et al., 2021). One reason for this is that consumers perceive micro-influencers to be more authentic than mega-influencers, which has an overflow effect on brand authenticity as well. However, it's worth noting that this positive effect of micro-influencers only works for "hedonic products" (e.g., chocolate) but not for "utilitarian products" (e.g., water bottles). Once again, the key takeaway here is that there is no one-size-fits-all strategy in influencer marketing, and that brands need to consider a wide range of situational factors when working with social media influencers.

When influencer marketing derails

You might have already heard about this—the infamous pop culture event FYRE Festival. The festival was an event promoted and hosted by American rapper JaRule and entrepreneur Billy McFarland in April 2017. Initially, the organizers intended to use the event to draw attention to the mobile application Fyre, which handles the booking process for those looking to hire a musical act or public figure to appear at a birthday party or event. Public promotion for the event began four months before the event was set to take place, and the company heavily engaged in influencer marketing. The event was sold as an exclusive event at an island in the Bahamas, with luxury-oriented lodging and the involvement of high-profile celebrities. And a number of well-known models were engaged in sharing promotional content related to the festival—often with no disclosure about their ties with the event or the nature of their posts.

However, the logistics for the event turned out to be more complicated than the organizers had initially anticipated. And with limited time to make appropriate changes and the organizers' unwillingness to cancel the event, the festival finally turned into a disastrous experience for the attendees, with scarce lodging, entertainment, transportation, and even basic food services. Due to this and other fraudulent activities, the organizer Billy McFarland was eventually charged by the federal prosecutor, which resulted in a sentence of 6 years in prison.

A detailed account of the event can be found in Netflix's *Fyre: The Greatest Party That Ever Happened*.

Key issues in social-media-marketing industry

Whenever profit is involved, problems are not far off. This is particularly true in areas such as social media marketing, where industry practices and technologies are constantly evolving. It's important to exercise prudence instead of recklessly pursuing maximum revenues. In the interest of brevity, this chapter outlines three main concerns related to three different stakeholders in the social media marketing industry: governments/policymakers, tech and social media companies, as well as influencers.

First and foremost, governments and regulating bodies across the world should consider appropriate measures to guide (and limit) social media marketing targeting vulnerable populations. (Readers might recall thorough discussions in Chapter 7, where we elaborated on how children and the elderly may fall victim to ill-intentioned social media practices.) In the context of social media marketing, more specifically, it has become a fact that children now spend more time on screen than ever, watching content that is finely-attuned to their cognitive developmental stages or produced by influencers of their age and background. Coupled with the fact that children often lack appropriate media literacy and are immature in critical thinking, it is therefore no surprise that young children and teenagers are uniquely susceptible to the commercial messages incorporated in social media content (De Veirman et al., 2019). These concerns led to the American Pediatrics Association (Radesky et al., 2020) recommending to "ban all commercial advertising to children younger than 7 years, and limit advertising to older children and teenagers" and to "prohibit in-app host selling and purchases, including loot boxes that pressure gamers to spend money during gameplay" (p. 6).

Another recurring issue that rises to the industry and government level concerns the collection, use, and reselling of social media users' data. In order to determine which influencers to work with, where to place ads, and which population to target their messages toward, social media platforms—particularly those that generate the majority of their revenue from advertising (such as Google, Facebook, and TikTok)—are racing to provide the most accurate, comprehensive, and live data to feed their commercial partners, to the extent that user data production can be neglected (see also Chapter 4).

To be sure, there are already some laws and regulations in place to protect user data. In the US, for instance, the FTC carries The Children's Online Privacy Protection Act (COPPA), which is designed to protect children under the age of 13 and "prohibits unfair or deceptive acts or practices in connection with the collection, use, and/or disclosure of personal information from and about children on the Internet" (Electronic Code of Federal Regulations, 2019). Unfortunately, though, the enforcement of those laws can be lax in practice, and social media companies can often manage to get away by paying a fine (Min, 2019). Therefore, strengthening the enforcement of existing laws to prohibit personal and location data collection is of crucial need. Companies can do their parts to stop this too. For example, in 2021, Apple rolled out a feature called App Tracking Transparency, allowing users to "Ask App Not to Track" on their phones. The idea is that, by doing so, users can actively prohibit other apps from tracking their users without their knowledge (Morrison, 2022). The bottom line is that companies can do more, but when would they choose to do so is yet another question.

Last but not least, social media influencers face a number of ethical questions related to authenticity, including the selection of brands to work with, the type of content to produce, and how to disclose relationships with brands. These issues, referred to as the "ethics of authenticity" by Wellman and colleagues (2020), may impact the personal brand of the influencer and the value they provide to their followers. Interestingly, government regulatory body seems to have caught up with the industry practice this time. In 2019, the FTC published an instruction conveniently titled "Disclosures 101 for Social Media Influencers," aiming to provide authoritative guidelines for influencer marketing. In a nutshell, the guidelines advise influencers to disclose their ties with brands (such as using #sponsored, #paid, or #ad; "This video is sponsored by brand X") and not to intentionally obscure those ties. But what influencers choose to do oftentimes has a lot to do with how the brands/companies approach influencer marketing. Therefore, the FTC guidelines are also meant to be followed by brands and companies, even when approaching influencer marketing. In enforcing these guidelines, for example, FTC went after Teami, a manufacturer of detox tea and other wellness products, alleging them of making "unsubstantiated claims" by collaborating with various Instagram influencers, including models, singers, actresses, and Realty TV stars. The company ultimately paid nearly $1 million to settle the case and set a precedent for the industry (Leskin, 2020).

Keywords: Electronic word-of-mouth (eWOM); earned media; owned media; paid media; CSR; UGC marketing; VR advertising; parasocial relationships; micro-influencers; mega-influencers; Product-Endorser fit; COPPA; ethics of authenticity.

References

Ariel, Y., & Avidar, R. (2015). Information, interactivity, and social media. *Atlantic Journal of Communication*, 23(1), 19–30. https://doi.org/10.1080/15456870.2015.972404

Berger, J. (2014). Word of mouth and interpersonal communication: A review and directions for future research. *Journal of Consumer Psychology*, 24(4), 586–607. https://doi.org/10.1016/j.jcps.2014.05.002

Brown, S. (2020). *Let's get virtual: Measuring virtual influencer's endorser effectiveness* [Doctoral dissertation, Texas A&M University]. https://hdl.handle.net/1969.1/192198

Burcher, N. (2012). *Paid, owned, earned: Maximising marketing returns in a socially connected world*. Kogan Page Publishers. https://doi.org/10.5860/choice.49-6974

Chang, H. L., Chou, Y. C., Wu, D. Y., & Wu, S. C. (2018). Will firm's marketing efforts on owned social media payoff? A quasi-experimental analysis of tourism products. *Decision Support Systems*, 107, 13–25. https://doi.org/10.1016/j.dss.2017.12.011

Chekima, B., Chekima, F. Z., & Adis, A. A. A. (2020). Social media influencer in advertising: The role of attractiveness, expertise and trustworthiness. *Journal of Economics and Business*, 3(4). https://doi.org/10.31014/aior.1992.03.04.298

Chu, S. C., Chen, H. T., & Gan, C. (2020). Consumers' engagement with corporate social responsibility (CSR) communication in social media: Evidence from China and the United States. *Journal of Business Research*, 110, 260–271. https://doi.org/10.1016/j.jbusres.2020.01.036

Colicev, A., Malshe, A., Pauwels, K., & O'Connor, P. (2018). Improving consumer mindset metrics and shareholder value through social media: The different roles of owned and earned media. *Journal of Marketing*, 82(1), 37–56. https://doi.org/10.1509/jm.16.0055

De Veirman, M., Hudders, L., & Nelson, M. R. (2019). What is influencer marketing and how does it target children? A review and direction for future research. *Frontiers in Psychology*, 10, 2685. https://doi.org/10.3389/fpsyg.2019.02685

Einwiller, S., Lis, B., Ruppel, C., & Sen, S. (2019). When CSR-based identification backfires: Testing the effects of CSR-related negative publicity. *Journal of Business Research*, 104, 1–13. https://doi.org/10.1016/j.jbusres.2019.06.036

Electronic Code of Federal Regulations. (2019). *Title 16: Commercial practices. Part 312- children's online privacy protection rules*. www.ecfr.gov/current/title-16/chapter-I/subchapter-C/part-312?toc=1

Erkan, I., & Evans, C. (2016). The influence of eWOM in social media on consumers' purchase intentions: An extended approach to information adoption. *Computers in Human Behavior*, 61, 47–55. https://doi.org/10.1016/j.chb.2016.03.003

Fernández, P., Hartmann, P., & Apaolaza, V. (2022). What drives CSR communication effectiveness on social media? A process-based theoretical framework and research agenda. *International Journal of Advertising*, *41*(3), 385–413. https://doi.org/10.1080/02650487.2021.1947016

Horton, D., & Wohl, R. R. (1956). Mass communication and parasocial interaction: Observations on intimacy at a distance. *Psychiatry*, *19*, 215–229. https://doi.org/10.1080/00332747.1956.11023049

Katz, E., & Lazarsfeld, P. (1955). *Personal Influence*. Free Press.

Kim, D. Y., & Kim, H. Y. (2021). Influencer advertising on social media: The multiple inference model on influencer-product congruence and sponsorship disclosure. *Journal of Business Research*, *130*, 405–415. https://doi.org/10.1016/j.jbusres.2020.02.020

Leskin, P. (2020). *Detox tea maker fined $1 million over 'deceptive' Instagram influencer ads claiming its tea could help you lose weight and fight cancer*. www.businessinsider.com/instagram-influencers-teami-detox-tea-sponsored-posts-ftc-settlement-2020-3

Lou, C., & Yuan, S. (2019). Influencer marketing: How message value and credibility affect consumer trust of branded content on social media. *Journal of Interactive Advertising*, *19*(1), 58–73. https://doi.org/10.1080/15252019.2018.1533501

Martínez-López, F. J., Anaya-Sánchez, R., Esteban-Millat, I., Torrez-Meruvia, H., D'Alessandro, S., & Miles, M. (2020). Influencer marketing: Brand control, commercial orientation and post credibility. *Journal of Marketing Management*, *36*(17–18), 1805–1831. https://doi.org/10.1080/0267257x.2020.1806906

Min, S. (2019). Google to pay $170 million for violating kids' privacy on YouTube. *CBS News*. www.cbsnews.com/news/ftc-fines-google-170-million-for-violating-childrens-privacy-on-youtube

Morrison, S. (2022). The winners and losers of Apple's anti-tracking feature. *VOX.com*. www.vox.com/recode/23045136/apple-app-tracking-transparency-privacy-ads

Müller, J., & Christandl, F. (2019). Content is king–but who is the king of kings? The effect of content marketing, sponsored content & user-generated content on brand responses. *Computers in Human Behavior*, *96*, 46–55. https://doi.org/10.1016/j.chb.2019.02.006

Nave, A., & Ferreira, J. (2019). Corporate social responsibility strategies: Past research and future challenges. *Corporate Social Responsibility and Environmental Management*, *26*(4), 885–901. https://doi.org/10.1002/csr.1729

Paramita, W., & Septianto, F. (2021). The benefits and pitfalls of humblebragging in social media advertising: The moderating role of the celebrity versus influencer. *International Journal of Advertising*, *40*(8), 1294–1319. https://doi.org/10.1080/02650487.2021.1981589

Park, J., Lee, J. M., Xiong, V. Y., Septianto, F., & Seo, Y. (2021). David and Goliath: When and why micro-influencers are more persuasive than mega-influencers. *Journal of Advertising*, *50*(5), 584–602. https://doi.org/10.1080/00913367.2021.1980470

Radesky, J., Chassiakos, Y. L. R., Ameenuddin, N., & Navsaria, D. (2020). Digital advertising to children. *Pediatrics*, *146*(1). https://doi.org/10.1542/peds.2020-1681

Reade, J. (2021). Keeping it raw on the 'gram: Authenticity, relatability and digital intimacy in fitness cultures on Instagram. *New Media & Society*, *23*(3), 535–553.

Santora, J. (2022). *Key influencer marketing statistics you need to know for 2022*. https://influencermarketinghub.com/influencer-marketing-statistics

Saxton, G. D., Gómez, L., Ngoh, Z., Lin, Y. P., & Dietrich, S. (2019). Do CSR messages resonate? Examining public reactions to firms' CSR efforts on social media. *Journal of business ethics*, *155*(2), 359–377. https://doi.org/10.1007/s10551-017-3464-z

Scholz, J., & Smith, A. N. (2016). Augmented reality: Designing immersive experiences that maximize consumer engagement. *Business Horizons*, *59*, 149–161. https://doi.org/10.1016/j.bushor.2015.10.003

Schouten, A. P., Janssen, L., & Verspaget, M. (2020). Celebrity vs. Influencer endorsements in advertising: The role of identification, credibility, and product-endorser fit. *International Journal of Advertising*, *39*(2), 258–281. https://doi.org/10.1080/02650487.2019.1634898

Wellman, M. L., Stoldt, R., Tully, M., & Ekdale, B. (2020). Ethics of authenticity: Social media influencers and the production of sponsored content. *Journal of Media Ethics*, *35*(2), 68–82. https://doi.org/10.1080/23736992.2020.1736078

Wolk, A. (2019). How William Wang and VIZIO are bringing TV into the 21st century. *Forbes.com*. www.forbes.com/sites/alanwolk/2019/06/27/how-william-wang-and-vizio-are-bringing-tv-into-the-21st-century/?sh=3e444dc843fe

12 Social media entertainment and well-being

Imagine you are a runner. Every day, before your routine jogging or walking, you take pleasure in selecting the most impeccably cared for and creatively designed running shoes in your possession. And while you are out, every step you take while running earns you money—not just any money but NFTs (non-fungible tokens, often seen as a form of virtual assets), which you can either cash out or invest in your shoes. Such a scenario may appear as though it is straight out of a dream, don't you think? This is STEPN, one of the first "move2earn" NFT mobile games. (Note that this example should not be taken as an endorsement for NFTs and related industries.)

It is difficult to predict the future of a startup company like STEPN, but examining the nature of platforms like it remains a meaningful task. Some consider it a social game, where sociality is embedded in the buying and trading of virtual running shoes. Others see it as more akin to a fitness app, which facilitates users to engage in physical exercise. For prudent observers, it might just be a variant of the all-too-complex virtual asset investment tool disguised as a game. Regardless, one thing we could perhaps agree on is that it is fun and entertaining.

Types of social media entertainment

The world of social media entertainment (SME) is incredibly diverse, encompassing everything from adorable animal pictures and witty memes to glamorous YouTube videos and cutting-edge VR experiences. The sheer range of SME content is staggering, and it's nearly impossible to capture its many forms in a single list. Nonetheless, we can begin to identify different types of SME content based on how users interact with them. Because social media entertainment is so inherently tied to the preferences and habits of its users, this categorization will inevitably be somewhat fluid and imprecise, but it may still provide some useful insights into the world of SME.

User-generated videos

One of the main types of social media entertainment is user-generated videos, which consist of audio-visual content produced by various users on different platforms, both professional and non-professional. These videos are consumed in a manner jibe with traditional television programs, through viewing. Within this category, there are three main genres: gameplay, DIY beauty, and personality vlogging (Cunningham & Craig, 2017). Gameplay refers to videos created by gamers discussing their gameplay, which may be recorded live or streamed live so viewers can see the player's spontaneous actions and reactions during the game. Gameplay is the most popular type of content on social media, with nearly 9 billion hours viewed in the first quarter of 2021 (Statisca, 2021). On YouTube, gaming content creators such as PewDiePie (Felix Kjellberg), ninja (Richard Tyler Blevins), and elrubiusOMG (Rubén Doblas Gundersen) are among the most subscribed channels, with many of them being non-US based young males.

In contrast, DIY beauty videos are created mostly by female creators, with content creators discussing topics related to cosmetics, fashion, and living style, among others. Not surprisingly, creators in this category often collaborate with companies for social media marketing (see Chapter 11), and as their accounts grow, many tend to launch their own beauty and fashion brands. Some popular DIY beauty YouTubers include Huda Kattan, Michelle Phan, and NikkieTutorials, who have all amassed millions of subscribers on their channels by sharing makeup tutorials, skincare routines, and fashion advice with their audiences.

Finally, personality vlogging includes the ones that offer comments and opinions on a wide range of cultural, civic, and political issues. Viewers tend to be attracted by the distinctive personalities of these creators. In that realm, readers may have encountered Hank Green (vlogbrothers), Natalie Wynn (ContraPoints), Jordan Peterson (JordanBPeterson), and Cenk Kadir Uygur (TheYoungTurks). Granted, there are many other popular and emerging genres of user-generated videos, such as life hacks, cooking instructions, education, expensive stunts/challenges, autonomous sensory meridian response (ASMR) videos, and product unboxing/testing videos. Therefore, what we consider popular genres today might change as users keep producing and propagating new types of content.

Social network sites

The second form of SME involves user interactions on social networking sites (e.g., TikTok, Instagram, and Snapchat), which entail behaviors such as posting, browsing, sharing, liking, and commenting. Unlike watching user-generated videos, where viewers tend to take a secondary role and their reactions are somewhat passive, interactions on social networking sites demand active participation from users (even passive browsing of one's news feed on Twitter consists of the active behavior of scrolling).

Among the various types of SNSs, social Q&A sites, such as Reddit, Quora, Yahoo Answers, and Zhihu, are worth mentioning. Social Q&A sites refer to online platforms that provide virtual spaces where users can propose questions, seek information, and build communities around participation. While you may have seen people conducting Q&As on YouTube and Facebook, questions and responses on social Q&A sites are directly generated and dynamically updated by a group of voluntary users (Shah et al., 2009). Moreover, the sheer variety of questions and topics covered on these sites is immense, ranging from mundane ones like "How to nail a job interview?" to more unusual and entertaining ones such as "How much water would it take to put out the sun?"

Social entertainment

The third type of SME is social entertainment, which consists of second screening (also known as Social TV), live commerce, social gaming, and VR/AR. As discussed briefly in Chapter 10, second screening refers to a shared viewing experience in which different media technologies and platforms are integrated into traditional television viewing. Behaviorally, it involves seeking and exchanging information through social network sites while watching live entertainment (or non-entertainment) programs (Ji, 2019). In recent years, the popularity of second screening has brought many audience analytics companies into the business of measuring audiences' social media engagement for various TV programs. Nielsen's Social Content Ratings represent an impactful attempt. Like social TV, live commerce and social gaming take social media engagement to the context of shopping and digital gaming. While live commerce is still an emerging practice in the US, social gaming has become a global phenomenon. Prominent examples include competitive MOBAs (Multiplayer Online Battle Arena) such as *Fortnite* and *League of Legends*, as well as casual games like *Animal Crossing* and *Minecraft*. Finally, there are virtual reality (VR) and augmented reality (AR), which refer to a range of technologies and devices that aim to provide a simulated experience that can be similar to or completely different from the real world. Among myriad examples, you may have heard of VRChat, a virtual social and chatting platform designed for virtual reality headsets such as Meta Quest, or *Pokémon Go*, a mobile social game that combines location-based and AR technology to facilitate players visiting real-world locations to capture virtual creatures.

Figure 12.1 Virtual Reality as a new frontier of social entertainment
(*Source*: Travelpixs/Shutterstock)

Live commerce

The earliest form of live commerce can be traced back to 1949, when William G. "Papa" Barnard created the first infomercial in the US to demonstrate how his blender could help families eat healthier. In the 1980s, TV shopping became popularized, with specialty shopping channels or networks being established (e.g., the Home Shopping Network) and gaining huge traction in the United States. However, as the Internet and social media took off, TV shopping morphed into live commerce or live-stream e-commerce.

According to the consulting company McKinsey (2021), the e-commerce company Alibaba first popularized live commerce in China. Today, it is a commonly accepted way of shopping for people in many Asian countries such as South Korea, Japan, and Thailand. Live commerce combines live-stream broadcast with e-commerce, making it similar to traditional TV shopping programs. However, what sets live commerce apart from its traditional counterparts is that it permits live exchange between viewers and the hosts: viewers could ask questions and expect to receive answers about the products immediately from the hosts, facilitating snap shopping decisions. Hosts are encouraged to develop their unique persona in this process, thus building a quasi-celebrity-fan relationship and making their argument to purchase more convincing.

Already, some US companies such as Amazon, Meta, and Wal-Mart have either launched or are in the process of pilot testing their own live commerce channels. Although the future of these channels remains to be seen, one thing is clear: some users enjoy watching them. Have you heard about live commerce before? If not, find and watch some live commerce content. Do you consider it a good form of social media entertainment? To what extent can cultural differences explain live commerce's popularity in Asian countries? Do you think this form of entertainment could sustain in the long run?

In sum, the three types of SME described earlier dominate the landscape of contemporary entertainment. However, these examples are far from comprehensive and are subject to change over time. As we look to the future, it is important to ask what exactly constitutes entertainment. Is there a unifying theme that runs through the seemingly infinite array of SME? Moreover, why is it that people of all backgrounds and persuasions can find commonality in the ways in which we use media for entertainment?

Defining entertainment

Fortunately, media psychologists have been examining these questions for decades. Many suggest that media entertainment, at its core, is a vehicle whereby people derive enjoyment. But what is enjoyment anyway? For most people, enjoyment refers to a hedonic feeling: a sense of fun, joy, excitement, relaxation, or positivity. That is perhaps what we perceive while watching TikTokers making goofy moves or twitch streamers scoring epic wins. However, the pleasurable aspect of enjoyment only captures one dimension of entertainment. In their often-cited writing, entertainment psychologists Zillmann and Bryant (1994) defined media entertainment as "any activity designed to delight and to a smaller degree enlighten through the exhibition of the fortunes or misfortune of others, but also through the display of special skills by other and/or self" (p. 438). Herein, the scholars highlighted the fact that media entertainment not only amuses us but also bears the power to broaden us.

The idea that entertainment can bring people to contemplation is an ancient one. A Shakespearean tragedy such as *Hamlet* and *King Lear*, for instance, has been widely circulated for centuries and often lauded as one of the highest forms of theatrical entertainment, a déclassé. Similarly, modern cinematic industries are accustomed to producing tearjerkers—movies such as *Spirited Away, The Notebook, Forrest Gump*, and *Call me by Your Name*, which often give the audiences a good cry, an empathic sadness, or a sense of meaningfulness. Considering this, Oliver and Raney (2011) borrowed the Greek term *eudaimonia* (loosely translated as happiness) and furthered their conceptualization of enjoyment as eudaimonic experiences, emphasizing that certain media entertainment may generate a host of mixed feelings, such as empathy, contemplation, and meaningfulness.

But those eudaimonic entertainments don't just make audience members think of themselves; some content is so unique that, upon exposure, one tends to forget about the notion of self and start to look outward. For instance, many types of music, especially the ones that feature unexpected harmonies, sudden dynamics, or the introduction of new voices, tend to give listeners goosebumps and experience a lump in the throat, both of which have been associated with awe—a strong emotion that is distinctively outward-focused (Ji et al., 2021). To capture this unique type of eudaimonic entertainment experience—feelings that orient audience members towards others, humanity's interconnectedness, and moral beauty—scholars coined the term self-transcendent media experiences or inspirational experiences (Oliver et al., 2018).

Exercise

Have you ever come across social media content that has been inspirational to you? Content that made you feel a sense of motivation, hope, or a desire to connect with others? If you are an active social media user, chances are that you have encountered quite a few of these types of content. According to a nationally representative survey, 53% of respondents reported having inspiring experiences while using social media (Raney et al., 2018). Now, try searching for some inspiring videos on YouTube or TikTok and watch them closely. Can you observe any patterns? Imagine yourself as a content creator. Can you identify common elements that elicit these inspirational feelings in people? Once you have made your list, refer to the article by Dale and colleagues (2017) in the chapter's reference list to see how your results overlap or differ from theirs.

How users enjoy social media entertainment

As a whole, the earlier tripartite conceptualization of media entertainment—as hedonic, eudaimonic, and self-transcendent experiences—might explain how we derive enjoyment out of most aforementioned SME. Because, just on the content facet, all SME, regardless of length or media format—a quick emoji exchanged via Snapchat, a stint at the VR world, or a lengthy vlog circulated on YouTube—lead to a combination of pleasure, appreciation, and inspiration. As platform users or content consumers, we tend to formulate a holistic impression toward these various forms of content on the receiving end, rather than carefully decipher which element of the source lead to excitement, contemplation, and at exactly what level. It was not until we shifted our roles and became a SME content creator (particularly the professional kinds) that the knowledge of what elicitors/elements may work in our desirable directions became pertinent.

Explaining how we enjoy SME through the content aspect is just one perspective, albeit a major one. The approach falls short in helping us understand how users might enjoy the various activities they partake in on social network sites and digital games. In that regard, we ought to turn our attention to another perspective, which considers media enjoyment as the satisfaction of three intrinsic human needs: autonomy (a sense of control), competence (a sense of mastery), and relatedness (a sense of social connection). This perspective was first introduced by Tamborini and colleagues (2010) in their intuitive discussion about how people enjoy video games. Given social network sites allow users to participate in content consumption (e.g., reading others' posts), content creation (e.g., posting something of their own), and social interaction (e.g., messaging others), it is not a stretch of the mind to say that these activities fulfill individuals' intrinsic needs (i.e., autonomy, competency, and relatedness) in various ways. For instance, the interactive social functions of SNSs may reduce perceived loneliness, increase social support, and satisfy relatedness needs. In the meantime, users could actively build positive social images of themselves and engage in comparison with others, satisfying their competence needs. Lastly, the autonomy needs were met as SNSs grant users a strong sense of control through various synchronous and asynchronous communication features. In fact, in a pioneering study regarding the role of needs satisfaction in social media use, Reinecke and colleagues (2014) found that, overall, the satisfaction of the three needs accounted for 24% of why people enjoy using Facebook, suggesting a pretty substantial explanatory power of media enjoyment as intrinsic needs satisfaction.

While it is apprehensible that the content and interactivity of SME would serve as sources of enjoyment, what about those "in-betweens"? In a highly circulated Ted talk delivered in 2015, filmmaker Chris Milk popularized the metaphor of Virtual Reality as an "ultimate empathy machine," highlighting the technology's ability to change participants' perspective through virtual embodiment. So, what makes technologies like VR and AR unique in the entertainment industry? According to entertainment psychologists, it is the ability of VR to create a feeling of presence—a state where users forget they're in a virtual world and fully engage in the artificial environment (Lee, 2004). The stronger the sense of presence, the more authentic and spontaneous the user's reactions become. While, to an outsider, a user with a Head Mounted Display and handheld controllers may look like a conventional gamer, VR allows users to become fully immersed in a world where they can interact with their surroundings in ways that feel real. Studies have shown that VR can enhance users' hedonic feelings, empathic state, and altruistic and prosocial motivations (c.f. Rueda & Lara, 2020). Immersion is a crucial aspect of VR, and it is defined as the system's ability to create a convincing environment for users to interact with (Sánchez-Vives & Slater, 2005). VR systems achieve immersion through various features, such as image quality, sound quality, field of view, display resolution, and headset size and weight. As such, different VR systems offer varying levels of immersion, which affects users' experiences and acceptance of the technology.

From entertainment to well-being

It is evident that what we consume and how we consume social media entertainment matters. Entertainment is not just a means to ward off boredom but a way of life. In exploring the connection between leisure and happiness, Aristotle (trans. 1984) stated that "it [leisure] is better than occupation and is its end; and therefore, the question must be asked, what ought we to do when at leisure?" Bringing this matter to the contemporary era, as humanity enters the era of being permanently online and connected (Vorderer et al., 2017), it raises the question of whether the integration of social media into our lives is beneficial to our collective well-being.

Undoubtedly, the question of whether social media impacts well-being remains a hotly contested issue in society and popular culture. Incidences such as the Facebook Documents, congressional hearings regarding social media and mental health, and public debates over regulating social media giants for public good draw crusades against the apps and gadgets we use daily. Thus, where does science stand on this matter?

Before we respond to that pertinent question, we need to first address what the term "well-being" means. In a colloquial sense, well-being is often used interchangeably with words such as wellness, happiness, and pleasure. Because of the broad appeal in its meaning, people tend to emphasize different dimensions while invoking the term, such as physical well-being, mental well-being, financial well-being, and social well-being. In academic literature, theories of well-being are in no short supply too. Nevertheless, scholars tend to agree on two general types of well-being; namely, the cognitive well-being and the affective well-being. Cognitive well-being stresses one's holistic evaluation of one's living conditions; as such, it is often equated with the notion of satisfaction with life. In cross-sectional survey and other types of formal research, life satisfaction is often measured by asking people to rate on a battery of items such as "In most ways my life is close to my ideal" and "So far I have gotten the important things I want in life," which constitutes The Satisfaction with Life Scale (SWLS; Diener et al., 1985). To date, the SWLS remains one of the most influential ways of measuring subjective well-being, and you may find the scale in varying length and language (see http://labs.psychology.illinois.edu/~ediener/SWLS.html).

Affective well-being refers to the various emotions that we may experience on a day-to-day or moment-to-moment basis. It is suggested that frequently experiencing positive emotions and infrequently experiencing negative emotions indicates a positively attuned state of well-being, and vice versa. To assess affective well-being in research, people are often asked to respond with yes or no to questions such as "Did you experience the following feelings during a lot of the day yesterday? How about _____?" with each of several emotions, including sadness, anger, disappointment, worry, joy, and pride being reported separately. The average of these discrete emotions determines one's positive and negative scores (Watson et al., 1988). It is also important to note that cognitive and affective well-being are not the

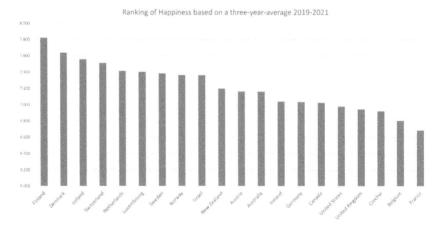

Figure 12.2 A list of top "happiness" countries in 2022
(*Source*: Helliwell et al., 2022)

only approaches scholars take to examine well-being. Depending on the research context, some may focus on social well-being, while others may examine ill-being, such as anxiety, loneliness, and depression.

The impact of social media on well-being

The scientific inquiry about social media's impact on well-being has been ongoing since the beginning of the Internet era. And the research methodologies used for tackling the problem have also evolved from cross-sectional ones (i.e., surveys and interviews), to longitudinal and, more recently, experimental ones. To ensure an apple-to-apple comparison, we will consider the evidence based on the types of methods used.

First, a vast majority of studies on this topic adopt survey methods, which rely on participants' self-report at the time of research rather than over an extended period. The findings were inconsistent, with results supporting both positive and negative associations. But meta-analyses show that, as a whole, the correlation between social media use and well-being is trivial and arguably negligible (Liu et al., 2019; Ferguson et al., 2021; Odgers & Jensen, 2020; Orben & Przybylski, 2019). The second group of studies used methods such as longitudinal designs and experience sampling (sampling participants' experiences multiple times a day/week) to provide data on both social media use and well-being over time and potentially probe the trends. Nonetheless, the results are no more clear-cut, with some suggested social media use associated with declines in well-being over time (Shakya & Christakis, 2017), and others revealed no connections (Vuorre et al., 2021) or limited effect contingent on many individual factors such as gender and age (Przybylski & Weinstein, 2017). Perhaps we could rely on experiments—the goldilocks for uncovering causal effects—to settle the argument? Not so easy! At the moment, the total number of experimental studies is rather small (albeit growing). Among these, most indicated a small but negative effect of social media usage (Kross et al., 2021), consistent with the minor correlational findings mentioned earlier.

So, where does this leave us? Should we conclude that social media is bad for us and that we should ditch it once and for all? Or, should we put forth laws and regulations to limit people's access to social media and smartphones? Not so fast—a couple of wrinkles about those finding might be worthy of our attention. First, in many of the formal research discussed earlier, researchers often consult indices such as statistical variance and effect size to gauge how *strong* one thing has to do with the other. In the case of social media and well-being, it is important to stress that many of these findings are extremely small (with one highly cited study showing social media used accounted for merely .04% of the well-being variance), and the effect sizes of various studies on the problem fall close to zero. In other words, the effects of social media on well-being are so small that many studies have found them to be almost negligible. On top of that, some argue that people's negativity bias may lead them to weigh their non-positive experiences with social media much heavier than the positive ones, especially when they are asked to recollect their memories during a formal research setting.

Clearly, the scientific debate on the topic is far from settled, and new evidence might update our knowledge to some extent too. For instance, there was experimental research uncovering that ditching social media (a.k.a. social media abstinence) for extended periods (1 to 4 weeks) *does not* improve people's well-being (Hall et al., 2019). Furthermore, in-depth interviews with those who had practiced social media abstinence or digital detox also raise the question of how long people can realistically sustain that effort. In fact, a large field experiment conducted in 2017 found that participants found Facebook to be more valuable after being restricted their access for 1 week (Mosquera et al., 2020). Therefore, simple measures such as restraining people's access appear overtly crude, and it is particularly so when weighing against the preponderance of benefits social media brought to us as discussed in previous chapters.

> ### The winkles of "social media use"
>
> One of the key issues underlying the debate about social media and well-being concerns how scholars conceive the notion of social media use. In most quantitative studies (e.g., surveys and experiments), researchers tend to ask participants to self-report their social media use by hours or minutes; per-day or per-week. This prompts some people to ask whether obtaining more objective measures (like the "screen time" auto-logged by your phone) would change the conclusion of the debate.
>
> To answer this question, Johannes and colleagues (2021) invited 96 participants to provide both self-reported and iPhone-logged social media use over the course of 5 days. And understandably, the difference between the self-reported measure and logged measure would indicate the accuracy of people's subjective evaluation. The researchers found that there was a systematic tendency for people to overestimate their daily social media use. But that tendency did not seem to matter when correlated with people's well-being: well-being was associated with neither self-reported social media use nor logged social media use. Concerning what predicts social media use, none of the measures for people's psychological traits (e.g., Big Five personalities) had anything to do with social media use, regardless

of how it was measured. What's more surprising is that even daily states (such as how motivational you feel and whether or not you are bored) are not related to social media use and accuracy.

Nonetheless, provocative ideas exist. In behavior genetics, researchers often study the impact of genes on behaviors by examining differences or similarities in people's behaviors between identical and fraternal twins. The basic idea is that identical and fraternal twins are all raised in a consistent household environment; thus, any behavioral difference observed between these two types of twins can be largely attributed to gene differences. Using a similar method, York (2017) found that nearly one- to two-thirds of the variance in social media use is explained by genetic traits. However, the result presented in this study is far from definitive due to its sample, measurement, and analytic methods. Thus, the jury is still out regarding genetics' impact on social media use. And finally, given what we know about the influence of social media use on well-being (which is very small), whether gene affects' screen time may not be directly relevant to the current debate after all.

Then how do we make sense of the converging views of the literature indicating a small but significant negative relationship between social media use and well-being? By now, you should understand that this very question in itself is a little too coarse. They meant to provide a brief sketch of the problem rather than informing individual-level practices. Of course, social media use will impact our emotional lives, or else we would not lament about phubbing, FOMO, and moral outrage online at all. Therefore, the more productive questions are *how we use social media*, *through what platform*, and *in what specific contexts*.

Fortunately, an increasing body of research in communication and media psychology is shedding light on these very questions. For example, we now understand that the likes, shares, and views we receive on social media, often at unpredictable intervals, trigger the release of dopamine, similar to the pattern of reinforcement that promotes habit formation, and in some cases, addiction, akin to slot machines, as some have argued. We also know that humans are social creatures, making us susceptible to social comparisons and emotional contagion. Studies have shown that viewing strangers' positive posts on Instagram can make users who engage in social comparison less happy, whereas those who don't engage in such comparisons report higher positive affect after viewing positive posts than after viewing neutral or no posts (de Vries et al., 2018).

Moreover, as we discussed in Chapter 5, many social media activities are a blend of self-presentation and self-expression. Frequent social media posters tend to express themselves in an authentic and self-idealized manner. For instance, in a longitudinal experiment, researchers asked 90 college students to post authentic and idealized self-expressions on Facebook for a week each. Participants generally reported significantly higher positive affect and mood during the authentic week than during the idealized week (Bailey et al., 2020). Therefore, being true to oneself seems to can contribute to one's overall happiness.

Finally, the design of social media platforms matters as well. While Jack Dorsey's admission that the "like" button may have made Twitter an unhealthy place is a rare exception among social media owners, ample evidence has shown that design features can influence norms and behaviors that either promote or hinder users' well-being (Masciantonio et al., 2021; Waterloo et al., 2018).

To sum up, the relationship between social media use and personal happiness is complex and dynamic, so seeking a definitive answer is misleading. However, social media designers, policymakers, and users all need to recognize that to harness the potential of social media for well-being and reduce its detrimental effects on our mental health, self-esteem, and moment-to-moment affect, we need to explore how we interact with these communication technologies, our motivations and psychology, and the context in which all activities occur. While there may never be a definitive conclusion on this matter, engaging in scientifically informed conversations about social media, as you have done by reading this book, is already a meaningful contribution to your well-being.

In conclusion, the relationship between social media use and personal happiness is complex and continually evolving, defying any definitive answer. However, it is crucial for social media designers, policymakers, and users to recognize that unlocking the potential benefits of these technologies for well-being and reducing their negative effects on mental health, self-esteem, and emotions requires examining our interactions with them, as well as the underlying motivations and contexts shaping these interactions.

While we may never arrive at a definitive conclusion on this matter, engaging in scientifically informed conversations about social media, as you have done by reading this book, is already a meaningful contribution to your well-being. Ultimately, the key to a healthier relationship with social media lies in ongoing self-reflection, critical thinking, and a willingness to adapt and evolve with the ever-changing landscape of digital technology.

Keywords: User-generated videos, social Q&A sites, social entertainment, live commerce, Virtual Reality, presence, eudaimonia, inspiration, intrinsic needs satisfaction, well-being, satisfaction with life, social media use.

References

Aristotle, J. B. (1984). *The complete works of Aristotle* (Vol. 2, p. 1984). Princeton University Press.

Bailey, E. R., Matz, S. C., Youyou, W., & Iyengar, S. S. (2020). Authentic self-expression on social media is associated with greater subjective well-being. *Nature Communications*, *11*(1), 1–9. https://doi.org/10.1038/s41467-020-18539-w

Cunningham, S., & Craig, D. (2017). Being 'really real' on YouTube: Authenticity, community and brand culture in social media entertainment. *Media International Australia*, *164*(1), 71–81. https://doi.org/10.1177/1329878x17709098

Dale, K. R., Raney, A. A., Janicke, S. H., Sanders, M. S., & Oliver, M. B. (2017). YouTube for good: A content analysis and examination of elicitors of self-transcendent media. *Journal of Communication*, *67*(6), 897–919. https://doi.org/10.1111/jcom.12333

de Vries, D. A., Möller, A. M., Wieringa, M. S., Eigenraam, A. W., & Hamelink, K. (2018). Social comparison as the thief of joy: Emotional consequences of viewing strangers' Instagram posts. *Media Psychology*, *21*(2), 222–245. https://doi.org/10.1080/15213269.2016.1267647

Diener, E. D., Emmons, R. A., Larsen, R. J., & Griffin, S. (1985). The satisfaction with life scale. *Journal of Personality Assessment*, *49*(1), 71–75. https://doi.org/10.1007/springerreference_184631

Ferguson, C. J., Kaye, L. K., Branley-Bell, D., Markey, P., Ivory, J. D., Klisanin, D., Elson, M., Smyth, M., Hogg, J. L., McDonnell, D., Nichols, D., Siddiqui, S., Gregerson, M., & Wilson, J. (2021). Like this meta-analysis: Screen media and mental health. *Professional Psychology: Research and Practice*. https://doi.org/10.1037/pro0000426

Hall, J. A., Johnson, R. M., & Ross, E. M. (2019). Where does the time go? An experimental test of what social media displaces and displaced activities' associations with affective well-being and quality of day. *New Media & Society*, *21*(3), 674–692. https://doi.org/10.1177/1461444818804775

Helliwell, J., Wang, S., Huang, H., & Norton, M. (2022). *Happiness, benevolence, and trust during COVID-19 and beyond*. https://worldhappiness.report/ed/2022/happiness-benevolence-and-trust-during-covid-19-and-beyond/

Ji, Q. (2019). Exploring the motivations for live posting during entertainment television viewing. *Atlantic Journal of Communication*, *27*(3), 169–182.

Ji, Q., Janicke-Bowles, S. H., De Leeuw, R. N., & Oliver, M. B. (2021). The melody to inspiration: The effects of awe-eliciting music on approach motivation and positive well-being. *Media Psychology*, *24*(3), 305–331.

Johannes, N., Nguyen, T. V., Weinstein, N., & Przybylski, A. K. (2021). Objective, subjective, and accurate reporting of social media use: No evidence that daily social media use correlates with personality traits, motivational states, or well-being. *Technology, Mind, and Behavior*, *2*(2), 1–14.

Kross, E., Verduyn, P., Sheppes, G., Costello, C. K., Jonides, J., & Ybarra, O. (2021). Social media and well-being: Pitfalls, progress, and next steps. *Trends in Cognitive Sciences*, *25*(1), 55–66.

Lee, K. M. (2004). Presence, explicated. *Communication Theory*, *14*(1), 27–50.

Liu, D., Baumeister, R. F., Yang, C. C., & Hu, B. (2019). Digital communication media use and psychological well-being: A meta-analysis. *Journal of Computer-Mediated Communication*, 24(5), 259–273. https://doi.org/10.1093/jcmc/zmz013

Masciantonio, A., Bourguignon, D., Bouchat, P., Balty, M., & Rimé, B. (2021). Don't put all social network sites in one basket: Facebook, Instagram, Twitter, TikTok, and their relations with well-being during the COVID-19 pandemic. *PLoS One*, 16(3), e0248384. https://doi.org/10.1371/journal.pone.0248384

McKinsey. (2021). *It's showtime! How live commerce is transforming the shopping experience.* www.mckinsey.com/capabilities/mckinsey-digital/our-insights/its-showtime-how-live-commerce-is-transforming-the-shopping-experience

Mosquera, R., Odunowo, M., McNamara, T., Guo, X., & Petrie, R. (2020). The economic effects of Facebook. *Experimental Economics*, 23(2), 575–602. https://doi.org/10.1007/s10683-019-09625-y

Odgers, C. L., & Jensen, M. R. (2020). Annual research review: Adolescent mental health in the digital age: Facts, fears, and future directions. *Journal of Child Psychology and Psychiatry*, 61(3), 336–348. https://doi.org/10.1111/jcpp.13190

Oliver, M. B., & Raney, A. A. (2011). Entertainment as pleasurable and meaningful: Identifying hedonic and eudaimonic motivations for entertainment consumption. *Journal of Communication*, 61(5), 984–1004.

Oliver, M. B., Raney, A. A., Slater, M. D., Appel, M., Hartmann, T., Bartsch, A., Schneider, F. M., Janicke-Bowles, S. H., Krämer, N., Mares, M. L., Vorderer, P., Rieger, D., Dale, K. R., & Das, E. (2018). Self-transcendent media experiences: Taking meaningful media to a higher level. *Journal of Communication*, 68(2), 380–389. https://doi.org/10.1093/joc/jqx020

Orben, A., & Przybylski, A. K. (2019). The association between adolescent well-being and digital technology use. *Nature Human Behaviour*, 3(2), 173–182. https://doi.org/10.1038/s41562-018-0506-1

Przybylski, A. K., & Weinstein, N. (2017). A large-scale test of the goldilocks hypothesis: Quantifying the relations between digital-screen use and the mental well-being of adolescents. *Psychological Science*, 28, 204–215. https://doi.org/10.1177/0956797616678438

Raney, A. A., Janicke, S. H., Oliver, M. B., Dale, K. R., Jones, R. P., & Cox, D. (2018). Profiling the audience for self-transcendent media: A national survey. *Mass Communication and Society*, 21(3), 296–319. https://doi.org/10.1080/15205436.2017.1413195

Reinecke, L., Vorderer, P., & Knop, K. (2014). Entertainment 2.0? The role of intrinsic and extrinsic need satisfaction for the enjoyment of Facebook use. *Journal of Communication*, 64, 417–438. http://dx.doi.org/10.1111/jcom.12099

Rueda, J., & Lara, F. (2020). Virtual reality and empathy enhancement: Ethical aspects. *Frontiers in Robotics and AI*, 160. https://doi.org/10.3389/frobt.2020.506984

Sánchez-Vives, M. V., & Slater, M. (2005). From presence to consciousness through virtual reality. *Nature Reviews Neuroscience*, 6, 332–339. https://doi.org/10.1038/nrn1651

Shah, C., Oh, S., & Oh, J. S. (2009). Research agenda for social Q&A. *Library & Information Science Research*, *31*(4), 205–209. https://doi.org/10.1016/j.lisr.2009.07.006

Shakya, H. B., & Christakis, N. A. (2017). Association of Facebook use with compromised well-being: A longitudinal study. *American Journal of Epidemiology*, *185*(3), 203–211. https://doi.org/10.1093/aje/kww189

Statisca. (2021). *Number of hours of video game live streams watched on streaming platforms worldwide in Q1 2019 to Q1 2021*. www.statista.com/statistics/1125469/video-game-stream-hours-watched/

Tamborini, R., Bowman, N. D., Eden, A., Grizzard, M., & Organ, A. (2010). Defining media enjoyment as the satisfaction of intrinsic needs. *Journal of Communication*, *60*, 758–777. http://dx.doi.org/10.1111/j.1460-2466.2010.01513.x

Vorderer, P., Hefner, D., Reinecke, L., & Klimmt, C. (2017). Permanently online and permanently connected: A new paradigm in communication research?. In *Permanently online, permanently connected* (pp. 3–9). Routledge. https://doi.org/10.4324/9781315276472-1

Vuorre, M., Orben, A., & Przybylski, A. K. (2021). There is no evidence that associations between adolescents' digital technology engagement and mental health problems have increased. *Clinical Psychological Science*, *9*(5), 823–835. https://doi.org/10.1177/2167702621994549

Waterloo, S. F., Baumgartner, S. E., Peter, J., & Valkenburg, P. M. (2018). Norms of online expressions of emotion: Comparing Facebook, Twitter, Instagram, and WhatsApp. *New Media & Society*, *20*(5), 1813–1831. https://doi.org/10.1177/1461444817707349

Watson, D., Clark, L. A., & Tellegen, A. (1988). Development and validation of brief measures of positive and negative affect: The PANAS scales. *Journal of Personality and Social Psychology*, *54*(6), 1063. https://doi.org/10.1037/0022-3514.54.6.1063

York, C. (2017). A regression approach to testing genetic influence on communication behavior: Social media use as an example. *Computers in Human Behavior*, *73*, 100–109. https://doi.org/10.1016/j.chb.2017.03.029

Zillmann, D., & Bryant, J. (1994). Entertainment as media effect. In J. Bryant & D. Zillmann (Eds.), *Media effects: Advances in theory and research* (pp. 437–461). Lawrence Erlbaum Associates.

Index

Note: Page numbers in *italics* indicate a figure on the corresponding page.

A/B testing 10
access, to digital information 21–22
accessibility, message 7, *7*
Active Print/Packaging 171
activism 151, 153
activism investment 151
actual self 73
addiction, to self-tracking 126
adolescents 27–28
advertising: algorithms and 53–55; children and 102; as news 139; *see also* marketing
affective well-being 192
affordance(s) 37–48, *38*; conversation *38*, 39; group *38*, 39; identity *38*, 38–39; non-work-related employee matters and 46–48; in organizational perspective 42–48, *44*; presence *38*, 39; relationship *38*, 39; reputation *38*, 39; sharing *38*, 39; social interactions and 40–42; transactional 41–42; users perspective on *38*, 38–40; in workplace 43–46, *44*
AI *see* Artificial Intelligence (AI)
algorithms: awareness of 56–57; children and 102; health information and 119; news and 137; social media 51–56, *54*; social media economy and 50–57, *51*, *54*

ALS Ice Bucket Challenge 140
Amazon 58
AMC Theatres 151
anonymity 74
APIs *see* application programming interfaces (APIs)
Apple 59, 180
application programming interfaces (APIs) 10
AR *see* augmented reality (AR)
Arab Spring 149, 156
Aristotle 191
ARPANET 3
Artificial Intelligence (AI) 51
(a)synchronicity 8, 74–75
attention crisis 62–66
augmented reality (AR) 171
authenticity 40–41, 159

badges 37
Bannon, Steve 160
Barnard, William G. 187
beauty videos 185
Berners-Lee, Tim 14
Biden, Joe 161
big data 122, 125–126
Big Five personality traits 23
BlackLivesMatter 150, 154
Bouazizi, Mohamed 149
boundaries 65
Brown, Michael 154
bullying 20, 26–27, *27*

Bush, Jeb 160
ByteDance 46, 196

California 106
callout/cancel culture 156
Cambridge Analytica 165
Canada 158
cancel culture 156–157
capitalism, knowing 126
Cerebral 121
children: advertising and 102, 179; algorithms and 102; developmental stages of 99–100; exploitation of 107; as influencers 102–106, *103*; laws and regulations with 105–106; video content oriented toward 101–102; virtual reality for 106–107; YouTube and 100–101
Children's Online Privacy Protection Act (COPPA) 180
China 156
clicktivism 155
Clinton, Bill 157
Clinton, Hillary 158–160, 164
communication characteristics, of social media 6–9, *7*
consignment fees 59
conspiracy theories *142*, 142–143
content: algorithms and 51–52; core issues about 15–16; democratization 15; moderation 52–53; news and 140; recommendations 51–52; in scope of social media 14; visibility and persistence 47
content creators, users as 13
content producers 60–62
control, retaking, from technology 64–66
conversation affordance *38*, 39
COPPA *see* Children's Online Privacy Protection Act (COPPA)
Corporate Social Responsibility (CSR) marketing 173–175
COVID-19 pandemic 42, 101, 111, 115, 117–120, 142, 164

Cruz, Ted 160, 165
CSR *see* Corporate Social Responsibility (CSR)
culture 16–17
cyberbullying *20*, 26–27, *28*
cyberslacking 47–48
cycle effect 155–156

data: meta- 13; private 13; public 13
Data Protection Authority (DPA) 11
data spectacle 123–124
Dean, Howard 157, 160
decentralization 14
DeGeneres, Ellen 80, *81*
delocalization 158
democratization 15
design, of social media: affordance and 37–40, *38*; as reactive 34
Design of Everyday Things, The (Norman) 37
developmental stages 99–100
Dickens, Charles 133
digital divide 20–22
digital immigrants 21
digital laborers 13
digital mourning 81
digital natives 21
Digital Services Act (DSA) 145
DingTalk 45
directionality, of relationships 86
Directive on the protection of young people at work (EU) 105–106
Direct-To-Consumer (DTC) Genomics 124–125
disinformation 138–143
dissolution, relationship 91–92
DIY beauty videos 185
Dorsey, Jack *36*
DPA *see* Data Protection Authority (DPA)
drugs, prescription 121
DSA *see* Digital Services Act (DSA)

echo chambers 137
Ecological Approach to Visual Perception, The (Gibson) 37
economy, social media: algorithms and 50–57, *51*, *54*; platform 57–62

Egypt 156
elections 157–165, *162*
electronic word-of-mouth (eWOM) 169
Elsagate 101
emoji 89–91, *91*
Enlightenment Now (Pinker) 3
enterprise social media 42–48, *44*
entertainment *see* social media entertainment (SME)
Epic Games 59
eudaimonia 189
eWOM *see* electronic word-of-mouth (eWOM)
exercise 122–123
exploitation: of children 107; health information and 118
expressive ties 95

Facebook 10, 31, 34, 75, *108*, 134, *135*, 144–145, 160
fact-checking 140–141
Fair Labor Standards Act (FLSA) 106
fake news 138–143
FCC *see* Federal Communication Commission (FCC)
Fear of Missing Out (FoMO) 24
Federal Communication Commission (FCC) 11
Federal Trade Commission (FTC) 11, 180
FFF *see* FridaysForFuture movement (FFF)
Fight Online Sex Trafficking Act (FOSTA) 12
filter bubbles 137
Finsta 76
fitness 122–123
fitspiration 122–123
Floyd, George 154
FLSA *see* Fair Labor Standards Act (FLSA)
FoMO *see* Fear of Missing Out (FoMO)
FOSTA *see* Fight Online Sex Trafficking Act (FOSTA)

fraud 110–111
FridaysForFuture movement (FFF) 152
FTC *see* Federal Trade Commission (FTC)
Fuchs, Christian 5
FYRE Festival 178–179

GameStop 151
GDPR *see* General Data Protection Regulation (GDPR)
gender, social media use and 28, 31
General Data Protection Regulation (GDPR) 11
genetic testing 124–125
Gibson, James 37
Gladwell, Malcolm 155

hacktivism 153
Harari, Yuval Noah 16
healthcare trends 119–122, *120*
health information: algorithms and 119; appeals of using social media for 116–117; consumer 115–116; defined 115; as emotional support 117; issues with seeking, via social media 117–119; overdiagnosis with 118; overreaction to 118; peer support and 117; privacy and 119; seeking 116; self-tracking and 123–126, *124*; as valuable data 125–126
Health Insurance Portability and Accountability Act (HIPAA) 119
Herman, John 37
HIPAA *see* Health Insurance Portability and Accountability Act (HIPAA)
homophily 15
Hong Kong 153

idealization, in online relationships 86–87, *87*
IF Metall 154–155
iGen 27–28

iGen (Twenge) 28
incidental exposure, to news 136–137
influencers 42, 102–106, *103*, 175–179, *178*
information brokers 15
information exclusivity 88
information-richness 74
Instagram 28, 34, *35*, 40–41, 76–77, *108*, *135*, 159–160
instrumental relationships 94–95
iPhone effect 29–30

Jackson, Michael 82

Kjellberg, Felix 53, 185
knowing capitalism 126
Kohlberg, Lawrence 99–100
Krieger, Mike 34

Lark 46
Leonsis, Ted 3
"likes" 34, *35*, 196
LinkedIn 76, *108*, *135*
literacy, technology 110
local journalism 66
lurkers 24–26

Machiavellianism 26
maintenance, relationship 88–89
marketing: children and 179; Corporate Social Responsibility 173–175; earned media in 172; forecast *170*; influencers and 175–179, *178*; key issues in 179–181; owned media in 172–174; paid media in 170–171; triad 170–175; user-generated content in 173; word-of-mouth 169; *see also* advertising
Marwick, Alice 143
McDonald's 173
McFarland, Billy 178–179
memes 161–162
memory: social media and 78–79; working 63–64

Meta 34
#MeToo 150, 152
Milk, Chris 190–191
misinformation 117–118, 138–143, *142*
modality, media 8
moderation, content 52–53
moral posturing 157
moral reasoning 99–100
motivations, of social media use 30–31
mourning 82
multitasking 63–64
Musk, Elon 13, 144–145, 172
Myanmar 144

neuroticism 23
New Hampshire 158
news algorithms 137
news exposure, incidental 136–137
news fabrication 138–139
news fact-checking 140–141
news use 134–137, *135*
Nextdoor *135*
NFTs 184
Norman, Donald Arthur 37
notifications 37, 65

Obama, Barack 81, 158, 160
Occupy Wall Street 153
O'Reilly, Tim 14
organizational perspective, affordance in 42–48, *44*

parasocial relationships 176
personality traits 23
personality vlogging 185
personalization, message 7, *7*
photographs 79–81, *81*
photo manipulation 139
phubbing 93–94
Piaget, Jean 99
Pinker, Steven 3
Pizzagate 158
platform economy 57–62
polarization 111
political campaigns 157–165, *162*

political communication 136–137
populism 164
posters, as user type 24–26
privacy 75–76, 119
professional relationships, social media in 94–95
protests 151–155
psychological traits, of users 22–24

QAnon 142
Quantified Self movement 124
Quora 14–15

reciprocity 89
Reddit 75–76, 125, *135*, 137, 151
red dots 37
relationship management: nature of relationships and 85–86; online relationship stages and 86–93, *87*, *91*; problematic social media use in 93–94; professional relationships and 94–95
relationships: development 87–88; directionality of 86; dissolution 91–92; intensity of 85–86; maintenance 88–89; nature of 85–86; parasocial 176; problematic use of social media in 93–94; stages of online 86–93, *87*, *91*; strength of 85–86
Roblox 107
Rospars, Joe 160
Russia 139, 154, 158
Ryan Haight Online Pharmacy Consumer Protection Act 121

Sanders, Bernie 160–161
Satisfaction with Life Scale (SWLS) 192
scams 110
scope, of social media 9–16
Section 230 11–12
self 73; -disclosure 72–77, 87–88; -expression 72, 74–75, 196; -preferencing conduct 58;
-presentation 72–77, 196; -tracking 123–126, *124*
selfie 79–81, *81*
sensory perception 99
service providers 12
SixDegrees.com 3
slacking 47–48
slacktivism 155
SME *see* social media entertainment (SME)
Smith, Quintin 107
Snapchat 6, 28, 31, 72, 77–78, *108*, *135*, 160
social interactions: platform affordances and 40–42
sociality 5
socialization, organizational 47
social media: communication characteristics of 6–9, *7*; defined 8; defining 3–17; early 3–4; memory and 78–79; for news use 134–137, *135*; origin of term 3; in professional relationships 94–95; scope of 9–16; self-expression and 74–75; timeline 4; typology of 5–6; well-being and 193–196
social media entertainment (SME): defining entertainment 188–189; social entertainment as 186–188; social network sites as 185–186; types of 184–188; user enjoyment of 190–191; user-generated videos as 185; well-being and 191–193
social movements: coordination of 152–153; defining 149–151; issues with 155–157; mobilization and 152–153; social media in support of 151–155
social penetration 87–88
software engineers 12–13
software engineers/developers 12
Spotify 54
stalking 92

stigmatized groups 78
stimulation 15
Sullivan, Maura 158
support groups 41
surveilling 92, 153–154
Sweden 154–155
SWLS *see* Satisfaction with Life Scale (SWLS)
symbolic thinking 99
synchronicity, media 8
(a)synchronicity 8, 74–75
Systrom, Kevin 34

Taiwan 153, 172
Tale of Two Cities, A (Dickens) 133
task-switching 63–64
technology 9–10; literacy 110
teens 27–28
Telegram 7, 139, 153
telehealth 119–121, *120*
Tesla 172
testing, genetic 124–125
TikTok 27–28, 41, 51–52, 61–62, *108*, 118, 123, *135*, 139, 152
town square 144
toxic apps 65
transparency 60
trolling *20*, 26–27, *28*
Trudeau, Justin 158
Trump, Donald 144, 159–160, 164–165
Tunisia 149
Twenge, Jean Marie 28
23andMe 125
Twitch 6, 31, *135*
Twitter/X 13, 19, *36*, 75, *108*, 134, *135*, 144, 152, 159, 162–163, 196

U&G *see* Uses and Gratifications (U&G)
UGC *see* user-generated content (UGC)
Ukraine 139, 154
user experience (UX) 9–10

user-generated content (UGC) marketing 173
user interface (UI) 9–10
users 19–31; affordances of social media in perspective of *38*, 38–40; as content creators 13, 60–62; demographics 21; lurkers 24–26; psychological traits of 22–24
Uses and Gratifications (U&G) 30–31
Uyghurs 156

VidCon 61
viral content 154
viral news 139–140
Virtual Reality (VR) 106–107, 184, 190–191
vlogging 185
VR *see* Virtual Reality (VR)

Walker, Paul 82
Wang, Alexandr 116
Web 2.0 14, 116
Web 3.0 14–15
WeChat 47
Weinstein, Harvey 157
well-being: from entertainment to 191–193; impact of social media on 193–196
WFH *see* working from home (WFH)
WhatsApp *108*, *135*, 152
#WhyIDidntReport 152
WikiLeaks 153, 158
word-of-mouth 169
working from home (WFH) 42–43
workplace: affordances of social media in 43–46, *44*
Wu, Tim 62

Yang, Andrew 158, *162*
YouTube 6, 28, 53, 59, *59*, 100–101, *108*, 134, *135*

Zoom 43

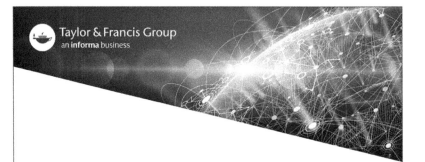